THE AMERICAN CIVIL WAR

A Military Overview

MICHAEL R. BRASHER

Coonewah Creek Publishing

To Caitlin and Benton

Preface

The American Civil War: A Military Overview resulted from my personal dissatisfaction with how most brief histories about the Civil War were written. Invariably they try to cram as much social, cultural, political and economic discussion into the narrative as possible while paying little attention to the military aspects of the war.

I decided if I had to pick just one facet of the war to try and tell a complete story in a reasonable number of words, it would have to be told from the military history perspective. That is what I have attempted to do in this narrative. Of course my approach is not meant to imply in any way that these other factors are not important to a complete understanding of the Civil War. Obviously they are, but I believed I could present an understandable perspective from the military viewpoint. I hope you will agree that I succeeded, at least so some extent, in that quest.

This is not a "scholarly" work that is heavily footnoted and includes a comprehensive bibliography. However you may rest assured that I have spent countless hours poring through primary and secondary sources of historical data from which to draw my own conclusions in writing this narrative.

If this book is successful, I hope to follow it up with additional works that will delve a bit deeper into some of specific military campaigns and battles and related aspects of our great American Civil War. I also plan to complete my "on again, off again" regimental history of the 2nd Mississippi Infantry Regiment and publish it in the very near future. I hope you will join me on my continuing journey.

"The Civil War defined us as what we are and it opened us to being what we became, good and bad things... It was the crossroads of our being, and it was a hell of a crossroads."

— SHELBY FOOTE

Introduction

With the election of the anti-slavery Republican candidate for President, Abraham Lincoln, the Southern states decided they had to take drastic action in order to protect their own interests. On December 20, 1860, a secession convention met in South Carolina and adopted an Ordinance of Secession from the Union. Mississippi, Florida, Alabama, Georgia, Louisiana, and Texas quickly followed suit. These states sent delegates to Montgomery, Alabama and on February 8, 1861, adopted a provisional constitution for the newly formed Confederate States of America. Jefferson Davis was chosen as the President for a six-year term of office.

The Constitution by which the permanent government of the Confederate States of America was formed was reported out by the committee and adopted by the Provisional Congress on March 11, 1861, to be submitted to the States for ratification. All States ratified it and conformed themselves to its requirements without delay. The Constitution varied in very few particulars from the Constitution of the United States, preserving carefully the fundamental principles of popular representative democracy and confederation of co-equal States.

These events were to set the stage for the bloodiest and saddest war in American history. In a conflict that combined elements of the Napoleonic Age with features of the new Machine Age, at least 600,000 Americans would lose their lives fighting for constitutional principle, sectional differences, economic self-interest, and moral righteousness. As a defining moment in United States history, our Civil War has no equal, which is why it remains such a fascinating subject even today.

Chapter One

The War Begins

The incident that began the Civil War involved the demand for the surrender of Fort Sumter, in Charleston, South Carolina harbor. On April 11, 1861, Brigadier General Pierre G. T. Beauregard formally requested that the fort be surrendered. The Federal commander, Major Robert Anderson, refused. At 4:30 a.m. on April 12, 1861, Captain George S. James fired the first shot of the war from a Confederate artillery battery. Artillery exchanges continued through April 13, when terms of capitulation were finally agreed to. The fort was evacuated by steamer at noon on April 14. The following day Lincoln issued a proclamation calling for 75,000 militia to serve for ninety days to put down "combinations too powerful to be suppressed" by the ordinary mechanism of government. The Civil War had begun.

The proclamation by Lincoln served to polarize the yet uncommitted states into action. Virginia, North Carolina, Arkansas, and Tennessee severed their ties with the Union, unwilling to supply troops to fight against their sister Southern states. The border states of Maryland, Missouri, and Kentucky, while providing soldiers to both armies, were kept under Federal control.

The "numbers" did not look good for the newly created Confederacy. Eleven states had left the Union; twenty-two remained. The population of the Confederate states was about nine million, almost one-third of whom were slaves. The Union states could count twenty-two million individuals and had a steady stream of immigrants. The South had only two main east-west railroad lines and limited ability to manufacture locomotives or rolling stock. Most of the known deposits of coal, iron ore and copper were in the North, together with about 92% of the country's industrial capacity. The Navy remained loyal to the Union and most of the merchant shipping was Northern-owned. If the South was to achieve victory, it would be against long odds.

Chapter Two

The Eastern Theater: "On To Richmond"

With Virginia having cast its lot with the South, the Confederate capital was soon moved from Montgomery, Alabama to Richmond, Virginia. This put the two opposing capitals, Washington, D.C., and Richmond, only 100 miles apart. This small area in Maryland and Virginia between the two capitals would see some of the bloodiest fighting of the war.

In the spring of 1861, Lincoln, seeing that his ninety-day volunteers' terms of enlistments would soon be expiring, placed Brigadier General Irvin McDowell at the head of the 30,000 men then in Washington and ordered an advance toward the Confederate capital. Although McDowell was unhappy with the untrained state of his troops, he proposed moving against Beauregard's concentration of about 22,000 Confederate troops near Manassas, Virginia. Delays in beginning the advance allowed Beauregard time to reinforce his position with some 9,000 troops under Brigadier General Joseph E. Johnston who had succeeded in giving a Federal "holding force" the slip, and moved his command by rail from the Shenandoah Valley to Manassas.

On July 21, 1861, a hot, dusty, Sunday afternoon, these two amateur armies clashed across Bull Run Creek. Although McDowell's attack plan was initially successful, a stubborn stand by Thomas J. "Stonewall" Jackson's brigade allowed Johnston's late arriving reinforcements to turn the tide for the Confederates. McDowell ordered a retreat, which soon became a rout. The inexperienced Confederates however, were in no shape to pursue the beaten Federals, and the Federal army, now more a disorganized mob, retreated back to Washington.

Chapter Three

❧

The Western Theater: The Opening Moves

When the other states initially began choosing sides, Kentucky, which was of strategic importance to the control of the Mississippi River, had declared its neutrality. Both sides hesitated to violate that neutrality by bringing in troops. Confederate Major General Leonidas Polk made the first move by occupying Columbus on September 4, 1861. Although Polk did not know it at the time, Grant had planned to occupy the city on the following day. Since the Confederates however, had been the first to violate Kentucky's neutrality, the state declared in favor of the Union.

Albert Sidney Johnston, who was regarded by many as the South's finest general, arrived to take command of the Western Department in mid-September, 1861. He could hardly have been pleased with the situation he found. He counted but 20,000 troops, most raw and ill-equipped, between the Appalachian Mountains to the east and the Mississippi River. In the Trans-Mississippi Theater, despite a

Confederate victory at Wilson's Creek, Missouri on August 10, 1861, Southern Generals Price and McCulloch exhibited a lack of cooperation which only vaguely suggested they were on the same side.

To correct these shortcomings, General Johnston immediately appealed for more troops and appointed Major General Earl Van Dorn as the ranking general over both Price and McCulloch as the new year of 1862 rolled in. Brigadier General Felix Zollicoffer was ordered to occupy the Cumberland Gap with a command of raw recruits to bolster Johnston's weak right flank.

Even with these measures however, the Federal forces opposing Johnston could have easily advanced right over his makeshift defenses. Johnston kept them from doing this by a combination of bluff and bluster. The theatrics were brilliant while they lasted but the first crack in Johnston's facade appeared on his right flank. It came about, not due to any Federal offensive movement, but because of General Zollicoffer's inexperience.

Brigadier General Crittenden was sent east to assume command of the right wing and found Zollicoffer was camped on the wrong (north) side of the unfordable Cumberland River. He was facing Brigadier General George H. Thomas' Federal command which was twice as large as his own. Crittenden ordered him to move back to the south bank, but in early January, Zollicoffer was still on the north side of the river. To compound problems, the Federal forces were starting to advance. Suddenly realizing his desperate circumstances, Zollicoffer launched a dawn attack on the Federal encampment at Mill Springs, Kentucky during a rain-soaked, dreary, January day. The attack failed and Zollicoffer was killed when he mistakenly rode into the Federal lines thinking the troops were his own men, although most of his command managed to escape to the south bank of the river.

Johnston's right flank had collapsed, but it did not prove to be his undoing. Thomas attempted to advance toward Nashville, but the barren nature of the region during the winter months stopped him about sixty miles short of his objective.

Johnston's cordon defensive line did not actually come unhinged until

February, 1862. However, when it did start to give way, the disintegration was, to say the least, spectacular. Johnston's Achilles' heel proved to be the combination of the Cumberland and Tennessee Rivers and their paths into vital Confederate territory. Although the Tennessee plunged deep into the heart of the Confederacy, Johnston's immediate concern was the Cumberland which curved past Clarksville, Tennessee (the site of the South's second largest ironworks) and Nashville, his base of supply. If Union gunboats were allowed to freely ply this river, his railway bridges would be quickly destroyed and his supply situation rendered untenable.

Fort Henry and Fort Donelson were built by the Confederates on the Tennessee and Cumberland, respectively, to block the rivers and prevent just this type of disaster. The forts were constructed at a point where the rivers were only twelve miles apart. However, all was not well. Fort Henry had been badly sited and was located on low ground subject to flooding and dominated by high ground across the river. Johnston's engineer had arrived in late November, 1861 and noticed the problem immediately. However, by mid-January, he had still not arrived at a solution.

Fort Henry fell to Flag Officer Foote's Federal ironclad fleet and the rising flood waters on February 6, 1862. Fort Donelson was sited much better, and Foote's fleet came out on the worse end of a heavy artillery duel with the fort. It was left to U.S. Grant's infantry to take Fort Donelson. The Confederate garrison attempted a breakout, and seemed to have been successful in opening the road that would allow their escape, but for some yet unexplained reason were ordered to fall back to their trenches. Thus about 15,000 badly needed Southern troops would become prisoners of war with the fall of Fort Donelson on February 16, 1862.

Johnston now knew he could no longer hold Nashville. He left Nathan Bedford Forrest in charge of the rear-guard to salvage what he could and fell back to Murfreesboro. Major General Buell, still advancing cautiously, did not reach the now undefended city of Nashville until February 23.

With the fall of Nashville, Major General Polk's position at Columbus, Kentucky was now untenable. He abandoned his fortifications and fell back. 7,000 Southern troops were sent to New Madrid and the fortress at Island #10 to block the Mississippi River. Another 10,000 were moved to the railway junction at Humboldt, Tennessee from which they could be rapidly shifted as the situation dictated.

In the Trans-Mississippi Theater, Confederate fortunes were also on the decline. General Van Dorn had been defeated on March 7-8, 1862, by Federal forces under Brigadier General Curtis at Pea Ridge, Arkansas, effectively losing the state of Missouri to Federal control.

Brigadier General Sibley's New Mexico campaign, aimed at gaining the eventual control of California and giving the Confederacy unblockaded access to Pacific ports, came to an abrupt end when he lost his wagon train following the fight at Glorieta on March 28, 1862.

New Madrid, on the Mississippi River, fell to Federal forces on March 13, 1862. Island #10 however, held out until April 7.

Meanwhile, on April 6, 1862, the largest battle on the North American continent up to that time was being fought near an unassuming West Tennessee Methodist meeting house called Shiloh Church. Grant's Army of the Tennessee was camped near Pittsburg Landing on the Tennessee River with about 48,000 men. He was awaiting the arrival of General Buell's Army of the Ohio with another 30,000 troops so they could combine and advance on the vital rail junction at Corinth, Mississippi.

Johnston and Beauregard, commanding the Confederate Army of the Mississippi, could not afford to wait until the two Federal forces united. With about 44,000 men, they advanced north from Corinth with the intent of striking Grant before Buell could join him. Ideally, Johnston would have preferred to await the arrival of Van Dorn's Trans-Mississippi command with another 15,000 men, but time was running out.

The Confederate attack on the morning of April 6 came as a complete surprise to Grant. Following the Confederate defeats at Forts Henry and Donelson, and their subsequent evacuation of Nashville, Grant had believed the Southern forces to be demoralized. He therefore did not bother to fortify his camps at Shiloh, nor did he send out adequate reconnaissance parties to warn of the impending Confederate attack.

The initial Southern attacks overran many of the Federal camps and rapidly pushed toward Pittsburg Landing. However, as the initial shock wore off, the veteran troops among the Federal forces began to stiffen their resistance. Prentiss' stand in the Sunken Road bought Grant the time he needed to patch together a final defensive line covering Pittsburg Landing. With the death of General Johnston and the resulting command confusion, the Confederates were not able to consolidate their forces for a final push against the Landing before evening.

That afternoon and night, the lead elements of the Army of the Ohio began to arrive and take position. The Federal armies began a general advance on the morning of April 7, and the now outnumbered Confederates were forced to withdraw back to Corinth. Casualties were about 13,000 Federals and 10,000 Confederates for the two-day fight.

Chapter Four

The Eastern Theater: The Peninsula Campaign

Following the Federal fiasco at First Manassas, Major General George B. McClellan replaced McDowell as commander of the Federal forces. He whipped the Federal Army of the Potomac into fine fighting trim, but was slow to move the army southward. After repeated urgings by the Lincoln government, McClellan finally decided to move against Richmond via the Yorktown Peninsula in March, 1862. However, it would be May before the troops actually saw any action.

From the Southern perspective, the defense of the Yorktown Peninsula was a major problem. Indeed, the overall Virginia Theater had a dismal outlook, with some 70,000 Confederates facing at least 200,000 Union troops.

The defense of the peninsula was handed to Major General John Bankhead Magruder, who, despite inadequate resources, set to work with enthusiasm. He had a long defensive line constructed with

Yorktown serving as its left flank. A secondary line was built some ten miles back from the first, just in front of Williamsburg. General Robert E. Lee, serving in an advisory capacity to President Davis at that time, was afraid these lines might be outflanked. On his advice, a third line was constructed about 10 miles in front of Richmond, with flanks anchored on the Chickahominy and James Rivers.

Magruder used his meager resources to their maximum effect, and by bluffing with the forces he had at hand, gave McClellan cause for hesitation in attacking. However, McClellan was having problems with his own government as well. On April 4, he learned that Fort Monroe, with a 12,000 man Federal garrison, had been taken from his command authority. Also, McDowell's 38,000 man corps would not be joining him on the Peninsula, but would be kept near Washington for its defense. Finally, he also learned that a stop had been put to additional Federal recruiting efforts.

Based on these distressing new developments, McClellan decided a siege was the solution. By early May he had set up 15 ten-gun batteries of 13" siege mortars. General Joe Johnston, now in command of Confederate field forces, did not want to face a bombardment by this heavy artillery, and ordered evacuation of Yorktown on May 3, leaving behind some 56 heavy siege guns of his own with ammunition, which McClellan added to his already plentiful supply. On May 5, the Confederate rear guard engaged the Federal advance elements in the battle of Williamsburg, and Johnston successfully pulled back even closer to Richmond.

Lee learned on May 16 that McDowell and some 40,000 Federal troops would be moving south toward the Confederate capital. This was a disaster in the making for the Confederacy. With McClellan poised to the east, opposed only by Johnston's forces, Johnston could not move to intercept McDowell without risking the capture of Richmond by McClellan. On the other hand, if Johnston did not move, then McDowell would take the city.

The solution to this problem lay in the character of President Abraham Lincoln. He was always incredibly concerned about any threat to his capital, and whenever he perceived such a threat Lincoln invariably overreacted. Thus Major General Thomas J. "Stonewall" Jackson was ordered to make an aggressive show in the Shenandoah Valley that would cause a perceived threat to Washington, D.C.

This sort of independent command suited Jackson very well. After engaging a Federal force at Front Royal on May 23, he then pushed General Nathanial Banks from his supply depot at Winchester on May 25. Besides netting some 3,000 prisoners, 9,000 small arms and tons of supplies, the operation had the desired effect of causing Lincoln to order McDowell to halt his advance on Richmond and to try and intercept Jackson. Actually, Lincoln may have reacted a little too well. He saw a chance to trap Jackson and so ordered Fremont's command from the west to take Harrisonburg and close the south end of the Shenandoah Valley.

Jackson however, did not "rattle" easily. He realized that his present location put him in a dangerous position. He was determined to make good his escape and take all the captured Federal booty with him. Although it was a close race, Jackson broke through the closing jaws of the Federal trap that resulted in a fight at Cross Keys on June 8 and a larger battle at Port Republic on June 9. With only 17,000 men, Jackson had neutralized the threat to Richmond of some 60,000 Federal troops.

Johnston could see that Jackson was performing splendidly. Now all he had to do was defeat the 100,000 or so Federal troops camped outside Richmond. Against this force, Johnston could bring some 70,000 Southern troops. The rain-swollen Chickahominy River offered a possible opportunity. McClellan, against his better judgment, had been ordered to split his own army across this river. Johnston saw that a rapid attack on the Federal wing on the south side of the river would give the Confederates a local numerical superiority and a good chance for success since Federal reinforcements could be brought in very slowly, at best.

The plan Johnston formulated was a good one, and relatively simple to execute. The divisions of Longstreet, Hill and Huger would advance east along parallel roads to attack Keyes in front of Seven Pines. However, the rain and mud caused everything that could go wrong to do just that. The march was disorganized and delayed. Men were drowned in crossing White Oak Swamp. At the end of the day, the Confederates could only claim the capture of 10 artillery pieces and 6,000 rifles. They had inflicted 5,000 casualties on the Federals, but had suffered 6,000 themselves. It could hardly be called an overwhelming victory. Moreover, Johnston was badly wounded and had to be relieved of command.

The man that took his place was none other than Robert E. Lee. Lee had an uncanny ability to "get into his opponent's head," and in the case of McClellan he saw that he would use his engineering expertise and superior fire-power to move slowly forward from one entrenched position to the next until he finally took Richmond. However, before Lee could cope with this, he needed time to improve his own defenses.

Fortunately for the Confederacy, the next ten days were continuous rain and McClellan's heavy artillery train was immobilized. The Confederates neutralized any attempt to bring them up by rail by their own 32-pounder artillery piece mounted on a railroad car — the first railroad gun in history. Shovels soon replaced muskets in the troops' hands and the earth was seen flying in the construction of new fortifications.

Lee pulled reinforcements from every quarter until he could muster an effective force of about 85,000 men. The new plan was to leave some 30,000 south of the Chickahominy in the newly-constructed entrenchments to hold McClellan's 75,000 on that side of the river and use the remaining 55,000 Southern troops to crush the 30,000 Federals on the north bank. If Lee were successful in defeating and destroying a large portion of this force, he would then capture McClellan's supply base and force him out into the open.

During June 12-15, J.E.B. Stuart's Southern cavalry rode completely around the Federal army, spreading confusion and confirming the Federal dispositions. Jackson was returning from his Shenandoah

Valley campaign and was due to arrive on June 25. To allow for possible delays, Lee planned the Confederate attack for June 26.

McClellan, for his part, was now convinced that he faced a massive army of some 200,000 Confederate troops and was badly outnumbered. If he really believed this was the situation, his subsequent actions during the confusing series of battles known as the Seven Days, become somewhat more understandable. To McClellan it seemed only logical that if Lee was attacking with 55,000 on the north bank of the Chickahominy, it must be a diversionary attack and the real blow would come in the south. The only prudent thing therefore, would be to fall back on the James River and Harrison's Landing.

The Confederates, after some minor fighting on June 25, moved north out of Richmond on June 26. Mechanicsville was taken easily, but an attempt to move east across Beaver Dam Creek was stopped by Federal forces in strong defensive positions. Jackson was supposed to have arrived and turned the flank of the position, but he did not show up that day.

The morning of June 27, the Beaver Dam Creek position was carried but only because the Federals had fallen back to another prepared position on Turkey Hill behind the Boatswain's Swamp Creek. Fitz-John Porter's command of 35,000 Federal troops was protected by a triple line of entrenchments with artillery support and marshy ground to their front. When Jackson's troops finally arrived that evening, the position was carried but with heavy Confederate casualties.

On Saturday, June 28, Lee spent much of the day trying to ascertain exactly where McClellan was retreating to. When Lee realized that McClellan was obviously falling back on the James River, he had to revise his earlier plans and decided to try and catch the Federals on either side of White Oak Swamp. The following day, Magruder was ordered to link up with Jackson and attack the retreating Federals. The Confederates were badly handled in a clash at Savage Station, primarily because Jackson again failed to show up on time. However, McClellan was forced to abandon much of his supplies, and an ammunition train sent forward to the Chickahominy railway bridge exploded with impressive results.

Monday, the sixth of the Seven Days, saw a remarkable lack of cooperation among the elements of the Confederate pursuit. Huger decided to cut an alternate road through the thick forest when he found his designated road blocked by felled trees. Holmes command ran into a naval bombardment. Jackson, who had difficulty in crossing the White Oak Swamp Creek, decided to lie down and take a nap at about 3:00 p.m.! As a result, only Longstreet's and A.P. Hill's troops were really involved in any fighting that resulted in another loss of some 3,300 Confederates at Glendale.

On July 1, the last of the Seven Days, Lee discovered that McClellan was protecting the last leg of his retreat by taking position on Malvern Hill. This defensive position was held by Porter and Keyes with two divisions each, more than one hundred artillery pieces, and a further four divisions in reserve. It looked formidable and it was. Lee first attempted to bring his artillery to bear on the position, but it soon became apparent that he was out-gunned. Lee looked for, and failed to find, a satisfactory alternative approach, but confusion in orders resulted in Huger, Magruder, and Hill launching a series of uncoordinated Confederate assaults. These resulted in nothing but another 5,500 Southern casualties. Jackson again failed to arrive in time to assist in the battle.

The series of hammer blows Lee had delivered during the Seven Days had achieved its objective of relieving Richmond from McClellan's forces. However, this had been accomplished at a very high cost. The Confederates lost 20,614 casualties compared to Federal losses of 15,849.

Chapter Five

The Eastern Theater: Second Manassas, Antietam and Fredericksburg

On June 26, 1862, a "Western" general, John Pope, received command of the newly created Federal Army of Virginia which was formed by the consolidation of the commands of McDowell, Banks, and Fremont (the Federal commands that Jackson had bested during his Shenandoah Valley Campaign). By July 12, this army had moved south to a point that threatened Richmond's access to the Shenandoah Valley.

Although McClellan was still a potential threat to Richmond, Lee felt he had to deal with this new development and sent reinforcements to Jackson with orders to "suppress" Pope. By August 3, McClellan was ordered to evacuate the Peninsula, thus removing the dual threat to Lee and allowing him the opportunity to concentrate on Pope exclusively.

The opposing armies maneuvered through mid-August, and by August

22 were facing each other across the Rappahannock River near Sulphur Springs, Virginia. Pope's lines were too strong for a frontal attack, so Lee directed forces around Pope's unsecured flanks to cut his supply lines. The first raid in Pope's rear, by Stuart's cavalry, failed in seriously damaging his supply line, but did net Pope's payroll of $350,000 (more than $8M in today's dollars) and captured the headquarters' copy of all the week's dispatches. The second raid was carried out by Jackson's "foot cavalry" and this time was much more successful. Jackson neutralized Bristoe Station and Manassas Station and destroyed Federal supplies that were distributed over almost a square mile.

Despite these setbacks, Pope still believed he was in a position to destroy Lee. Lee had split his army and the Federals were in position between Lee's two wings with superior forces to destroy either one. Pope hurried his command to Manassas hoping to smash Jackson, only to discover that he had apparently vanished into thin air. Eventually Jackson was discovered at Sudley Mountain near the now year-old battlefield of First Manassas.

Jackson grimly held his position against Pope's attacks through August 29. On August 30, Longstreet's wing of Lee's army arrived and turned the tide of battle. During the late afternoon, after Pope had committed his last reserves to the attack on Jackson, Longstreet launched a massive assault into the flank of the Federal army, routing it from the field.

Pope had indeed been "suppressed" at Second Manassas, losing some 16,000 casualties, 30 artillery pieces, 20,000 small arms and mountains of other supplies. Pope was relieved of command on September 2. McClellan was placed in command of all the forces around Washington, D.C.

Although Lee had gained another victory, it was unclear as to the best way to press his advantage. His forces could not stay in this area of northern Virginia, but to fall back would be to negate the advantages of his recent victory. His decision, therefore, was to invade Maryland. He hoped to gain support from the local populace of the state, and he also saw an opportunity to sway foreign opinion if he could win another victory on Northern soil. Washington, D.C. itself Lee knew was too strong to attack, but he hoped to be able to capture the 12,000 man Federal garrison at Harper's Ferry during his advance. To do so, he would have to take the risk to divide his army in enemy territory, but he felt that the Army of the Potomac was still demoralized from its recent defeats and McClellan, if remaining true to form, would react with all the speed of a tortoise.

The Army of the Potomac however, was not demoralized. It was to the contrary, still full of fight. Maryland did not welcome the Confederates with open arms, as had been hoped, and worst of all, McClellan had come by a copy of Lee's entire plan of operations for the Maryland Campaign. With this information in front of him, even McClellan was capable of moving fairly quickly.

Lee learned that McClellan had come into possession of a copy of his orders from an informer. His Army of Northern Virginia was now split into five segments. Lee desperately needed time to concentrate these elements to defend himself from the attack he knew McClellan would be planning.

It was a near-run thing, but D.H. Hill, with reinforcements from Longstreet, was able to hold McClellan's army at bay at Turner's Gap in South Mountain long enough for Lee to form a defensive position at Sharpsburg, Maryland, behind Antietam Creek on September 15, 1862. The 12,500 man garrison at Harper's Ferry surrendered that same day.

On September 16, McClellan was facing, at most, some 18,000 Confederates in line of battle. If he had attacked that day, he almost certainly would have crushed this small force. However, he did not attack, but instead spent the day planning and investigating the terrain; all this time, more and more Confederate reinforcements were arriving.

On September 17, McClellan finally attacked. Though Confederate reinforcements, in the form of the divisions of McLaws, Anderson, and A.P. Hill would be arriving throughout the day, McClellan at no time faced odds worse than two to one in his favor. Believing faulty intelligence estimates of Lee's strength, McClellan was unwilling to fully commit his army to the attack for fear of a Confederate trap. The result was a bloody see-saw battle that saw Lee and his outnumbered Confederates put in one of their best tactical performances and fight McClellan's army to a draw.

Lee retreated the following night and despite repeated urgings, McClellan failed to press forward a pursuit of Lee's forces. Antietam went down in history as the bloodiest single day of the war, with over 23,000 total casualties. Although a tactical failure for McClellan, he could count it a strategic victory since the Confederates had to retreat. This gave Lincoln the opportunity to issue his Emancipation Proclamation on September 23, 1862.

Despite continuing urging from Lincoln, McClellan could not be induced to advance and during the period of October 10-12, Stuart's cavalry again rode completely around the Federal army and did over $250,000 (almost $6M in today's dollars) in damages, causing the government much embarrassment. Lincoln had had enough — he fired McClellan and replaced him with Ambrose E. Burnside.

Major General Burnside did not want command of the Army of the Potomac, feeling that he was not competent to hold such a responsibility. He was ultimately overruled and forced to accept the assignment. To Burnside's credit, he did not suffer from false modesty — he was incompetent.

Burnside's offensive began well enough, stealing a march on Lee and moving rapidly down the Rappahannock River, planning to cross over to Fredericksburg on December 19, 1862. The pontoon bridges he had ordered, unfortunately, failed to arrive on time. This gave Lee the opportunity to shift his forces to cover the crossing. Excellent defensive terrain overlooked Fredericksburg on the south bank and Lee proceeded to construct prepared positions for his troops. Burnside should have seen that it was now an obvious mistake to attempt to assault Lee's lines. He did so anyway.

The results were as might be generally expected under these conditions. Federal troops were rapidly cut down before they even came near the Confederate positions. Luckily for the Federals, Lee did not have the strength to launch his own counterattack after the attack was repulsed. Union heavy artillery on the opposite bank of the river served to protect the retreat of the surviving Federals back across the Rappahannock.

In late January, 1863, Lincoln replaced Burnside with "Fighting Joe" Hooker to the command of the Army of the Potomac.

Chapter Six

The Western Theater: Bragg's Kentucky Campaign

With the fall of Corinth, Mississippi at the end of May, it was clear that Memphis could not be held. It fell on June 6 with the defeat of the Confederate ram fleet defending the city. The Mississippi River was now open to Federal gunboats as far south as Vicksburg, Mississippi.

The Federal command organization underwent major changes during this time period. The end of June, 1862, Major General Pope was ordered east to meet his fate at Second Manassas, relinquishing his command to Major General Rosecrans before departing. On July 11, 1862, Hallack was made commander-in-chief of all Federal armed forces, east and west, and went to Washington, D.C. Grant, who had seriously considered resigning following the close call at Shiloh, was appointed to overall command of Rosecrans' army and other forces in the theater, giving him control of some 75,000 troops.

Major General Buell's forces, having been ordered to march on Chattanooga, Tennessee, were having a rough time of it. With constant pressure from Washington and additional pressure in the form of Confederate cavalry and guerrilla attacks on his supply lines, his army slowly crawled forward on half rations. Then on August 12, 1862, John Hunt Morgan's cavalry destroyed an 800 foot-long tunnel on the Louisville & Nashville Railroad and cut off Buell from his base of supply at Louisville, Kentucky. This, combined with intelligence that Bragg was advancing north, led him to conclude he must fall back to protect Nashville.

General Braxton Bragg, now in command of Confederate forces in the theater, was determined not to simply stand on the defensive, but to go over to offensive operations to recover both Tennessee and Kentucky for the Confederacy. The campaign began favorably as Confederate forces in East Tennessee, under the control of General Kirby Smith and in cooperation with Bragg, moved north into Kentucky with 12,000 troops. At Richmond, Kentucky they met, on August 30, a command of 7,000 new Federal recruits defending the city. In a one-sided victory, Smith's casualties numbered only about 450 while the Federals lost 206 killed, 844 wounded, and 4,303 captured or missing. Lexington, Kentucky was captured by Smith's forces, unopposed, the following day.

On September 13, Bragg had reached Glasgow, Kentucky which placed him between Buell, now at Bowling Green, and Smith in Lexington. Bragg's forces moved north to the Green River and forced the surrender of another 4,000 man Federal garrison at Munfordville.

Buell advanced his forces again northward to Louisville, and then began a movement to the southeast towards Bragg's suspected location. The two armies eventually stumbled into each other outside Perryville, Kentucky on October 8, 1862. Bragg, who was outnumbered three-to-one, but did not think so at the time, ordered an attack by Hardee and Polk. This assault routed the Federal Left Wing under General McCook. On the opposite flank, Joe Wheeler's 1,200 Confederate cavalry managed to immobilize Crittenden's corps of 22,500 Federal troops in an impressive performance.

When the battle closed at the end of the day with no decisive results however, Bragg decided to retreat southward. Buell's pursuit was unenthusiastic, and Bragg arrived back in Knoxville on October 22. Lincoln was unhappy with the turn of events and on October 24, ordered Buell to turn over his command to Major General Rosecrans.

Meanwhile, further west, on September 20, Grant almost succeeded in trapping Major General Sterling Price's command of Trans-Mississippi troops at Iuka, Mississippi, about 20 miles east of Corinth. Again, with Confederate high command confusion, Price and Van Dorn could not decide whether to move their forces north, to link up with Bragg or to advance on some other objective. Van Dorn finally decided to assault Corinth, thinking it now only lightly defended. A total Confederate force of about 22,000 men advanced to the attack on October 3, 1862.

Corinth however, was not lightly defended. It contained an equal number of Federal troops and was encircled by a double ring of fortifications backed by artillery, the outer ring which had been built by the Confederates themselves before evacuating the town earlier that year. In some of the most vicious fighting of the war, the Confederate advance actually broke through to the town itself but were outflanked and a Union counterattack quickly drove them back out with heavy losses.

A continuation of the fight on October 4 produced no gains for the Confederates and Van Dorn ordered a retreat. Almost trapped by Rosecrans' pursuit and a converging Federal column ordered out by Grant, Van Dorn's forces suffered well over 4,000 casualties to Federal losses of 2,500. The western flank of Bragg's offensive campaign had suddenly collapsed.

After Rosecrans took command from Buell, Lincoln had expected some rapid offensive action to be taken. Despite repeated urging, Rosecrans found excuses to delay his movement from Nashville until almost Christmas. Then, having learned that one of Bragg's divisions had been detached to Vicksburg and that Forrest's and Morgan's cavalry commands were on raids elsewhere, he moved his Army of the Cumberland to the southeast. The march route was in three columns under Generals McCook, Thomas, and Crittenden. The opposing armies collided on December 31 just north of Murfreesboro at Stones River.

Both commanders had planned to attack the enemy right flank, but Bragg beat Rosecrans to the punch and the Federal army quickly had to go over to the defensive. Hardee's and Polk's Corps drove McCook's and Thomas' men back with a "hinge" at a point in the Federal line called the Round Forest. Rosecrans' line was bent at almost ninety degrees to his previous position by sunset but it had not broken. Apparently bolstered by Thomas' council at a meeting of his officers that night, Rosecrans decided not to retreat, even though he had suffered about 12,000 casualties to Bragg's 9,000.

Bragg was now faced with a dilemma. His remaining strength was really inadequate to force Rosecrans from his new position, so on New Year's Day, 1863, he waited, hoping Rosecrans would make the logical decision and retreat. When January 2 found him still in position, Bragg ordered an ill-advised assault on the Federal left flank by Breckinridge's division, which, despite some initial success, was blasted apart by the Federal massed artillery. Thus, January 2 only brought Bragg another 1,700 casualties which he could ill afford.

Finally conceding defeat, Bragg retreated during the night of January 3 back to Tullahoma, Tennessee. Rosecrans' army was too battered to attempt a pursuit, but Stones River could be claimed as a Federal victory due, more than anything else, to Rosecrans' stubbornness.

Chapter Seven

The Eastern Theater: The Chancellorsville Campaign

Lincoln appointed "Fighting Joe" Hooker to the command of the Army of the Potomac on January 25, 1863. Hooker immediately set out to improve the welfare and morale of the troops. He introduced corps insignia badges to give the men more pride in their units. He also reorganized the Federal cavalry into a single corps of 11,500 troopers under the command of Brigadier General George Stoneman to better counter the Southern cavalry superiority.

Hooker had more than 130,000 troops and 412 artillery pieces, more than twice the strength of Lee in all three combat arms: infantry, cavalry, and artillery. He intended to use these superior numbers to effect a plan whereby he would employ a pincer movement against Lee. One half of the army would cross the Rappahannock River below Fredericksburg and the other half would cross upstream to move against Lee's rear. Each Federal wing would be almost the size of Lee's entire command. The Federal cavalry meanwhile, would attempt to create confusion by operations behind Lee's lines.

Bad weather prevented the execution of the plan until late April. Hooker initially misled Lee as to his true intentions by leaving Gibbon's division in camp while moving the rest of the army. Lee quickly discerned Hookers's true intentions, however. After watching Sedgwick's men consolidating their bridgehead below Fredericksburg, Lee decided that the main threat was Hooker's flanking column. Lee therefore moved the bulk of his army towards Hooker and left Major General Jubal Early with about 10,000 men to contain Sedgwick.

For some reason on May 1, Hooker suddenly became cautious, halted his advance while still inside the tangle of the Wilderness and ordered his men into a defensive posture. Had he continued to more open country, his superior numbers would have given him a decided advantage, especially with respect to his artillery.

With Hooker paused in the Wilderness, Lee and Jackson conceived a bold but risky plan to strike Hooker first. Lee divided his forces once again and sent Jackson's corps on a long march to turn Hooker's unprepared right flank. Late in the afternoon of May 2, Jackson slammed into Hooker's flank, routing the XI Corps and apparently unnerving Hooker even more. Tragically for the Confederacy however, Jackson was accidentally shot by his own men during the confusing aftermath of the initial assault. He would die of pneumonia eight days later.

On May 3, Sedgwick attacked and broke through Early's defenses in an attempt to come to the aid of Hooker. He was stopped again near Bank's Ford and Salem Church. On May 4, Lee launched an attack against Sedgwick, but could not drive him from his position. Meanwhile, Hooker continued contracting his own lines and constructing defensive fortifications. However, later that night, his nerve apparently failing him again, Hooker ordered a full retreat of the Army of the Potomac.

Although Lee was very upset that Hooker's army had escaped, had he actually continued to assault the Federal army in its prepared defenses, he could very well have destroyed his own army instead.

Hooker was defeated more by his own loss of nerve than by Lee and Jackson. The troops of the Army of the Potomac were still full of fight, but "Fighting Joe" Hooker had had enough. He would be relieved of command of the army in mid-June.

Chapter Eight

The Western Theater: The Vicksburg Campaign

Vicksburg, a veritable fortress city, controlled the middle stretch of the Mississippi River. It was situated on high, unscalable bluffs from which its heavy artillery could control the river below. It was also very difficult to approach overland. Directly to the north was the vast Yazoo Delta, impassable to any large body of troops. Federal gunboats could not sail past Vicksburg without risking destruction. 100 miles downstream at Port Hudson, Louisiana a second Confederate river fortress prevented Union naval forces from moving upstream.

U.S. Grant initially planned a two-pronged advance on Vicksburg. Major General Sherman would move down the Mississippi River, turn the Federal fleet up the Yazoo River and attempt to land troops to the north of Vicksburg. In the meantime, Grant would move south down the railroad through Grenada, hopefully forcing Lieutenant General Pemberton's Confederate forces to move to try and intercept him.

This plan was a dismal failure, largely due to the efforts of Forrest's and Van Dorn's cavalry. Forrest commanded a raid in December, 1862 that cut Grant's supply lines in Tennessee to bits. Van Dorn, now seeming to have found his true calling as a commander of cavalry following the Corinth fiasco, was eager to redeem his reputation. During the same December time period, he led a second raid to destroy Grant's large supply depot at Holly Springs, Mississippi.

With his supplies destroyed, and thanks to Forrest, no way of bringing in more, Grant was forced to retreat to Memphis. Sherman likewise had no success in attempting to land troops at Chickasaw Bluffs and was repulsed with severe losses. Grant however, refused to be discouraged by these setbacks and kept at work toward his objective of Vicksburg. He would eventually make five more attempts and suffer five more failures before he achieved his final objective.

The first of the next five attempts was the effort to construct a canal across the tongue of land in front of Vicksburg to divert the river channel and bypass the city's artillery batteries. During this project, a dam gave way causing significant damage to the Federal camps and forcing the project to be scrapped.

The second involved trying to move the fleet via a circuitous route through Lake Providence, about 50 miles north of Vicksburg. The movement would involve traversing several bayous before rejoining the Mississippi River by way of the Red River a few miles above Port Hudson. The bayous were found to be too blocked by cypress trees and flood debris to carry out the plan.

The third project also was aimed at exploiting the complex river system around Vicksburg. This time an attempt would be made to move via a bayou called the Yazoo Pass, just south of Helena Arkansas. However, after initial success in moving into the Tallahatchie River system, further attempts were blocked by a Confederate fort and the Union fleet was forced back.

The fourth attempt would be through utilization of the Steele Bayou to reach Black Bayou, connecting with Deer Creek, connecting with Rolling Fork Bayou, connecting with the Sunflower River which flowed into the Yazoo River above Haines Bluff. Underwater vegetation fouled the Union fleet's paddle wheels, and they were forced to abandon this idea.

The final project involved the construction of a second canal just below Duckport that would allow the passage of light draft vessels, but falling river levels made this attempt impractical also.

Finally, although Grant was still no closer to his goal, the wet season had ended and Admiral Farragut had been successful in running the batteries at Port Hudson with two Federal gunboats and could now control the Mississippi below Vicksburg. Grant incorporated these new developments into his planning.

Grant requested and received agreement from Admiral Porter, to try and run his ships past the Vicksburg artillery batteries. Porter pointed out to Grant that if this attempt was successful and the ships were downstream of Vicksburg, there would be no way to move them back upstream while the Vicksburg batteries were still operational. On April 16 the plan was carried out on a dark, moonless night and was almost completely successful. Only one transport vessel was lost, and there were no casualties among the Federal personnel. A few nights later more army-owned transport vessels were run past Vicksburg with the loss of another transport and six of twelve supply barges.

To try and deceive the Confederates as to his true line of operations, Grant ordered diversions by Sherman at Haines Bluff and a cavalry raid by Colonel R.H. Grierson, which proved to be very successful in distracting Pemberton's attention.

Grant had intended to cross his troops from the west bank of the Mississippi River and land at Grand Gulf, Mississippi, but was unable to neutralize the Confederate forces at that place. The next day however, he moved the landing point to Bruinsburg and landed his forces without opposition on April 30.

Due to confusion caused by Grant's diversions resulting in a dispersion of the Confederate army, Pemberton was unable to rapidly field substantial forces to contain Grant's bridgehead. A Confederate task force under Brigadier General John Bowen met Grant's advance elements at Port Gibson on May 1, but after a tough all-day fight, was forced to retreat when no reinforcements could be sent by Pemberton.

Despite orders to cooperate in operations against Port Hudson, Grant now took it upon himself to implement a bold new operational strategy. Knowing that General Johnston was assembling an army in central Mississippi to come to Pemberton's support, Grant decided to rapidly move inland and interpose himself between the two Confederate forces, abandoning his river-based line of communications as he did so. He would instead have to live off the land until he could re-establish contact with the Union fleet. Pemberton meanwhile, despite urgings by Johnston to immediately concentrate his own forces and move against Grant's bridgehead, pulled the Confederate defenders back into the vicinity of Vicksburg.

On May 12, Grant began moving the Federal army of about 44,000 troops toward the interior of Mississippi. This force met and defeated a small task force under Brigadier General John Gregg at Raymond that same day and two days later broke up the concentration of Johnston's forces at Jackson. The Confederates abandoned the important railhead and supply depot and escaped to the north.

Again Johnston ordered Pemberton to move out of Vicksburg to strike Grant's rear and link up with his own relief forces. Delaying until May 15, Pemberton finally assembled three of his five divisions and marched out of Vicksburg with the intention of cutting Grant's now non-existent supply line. Changing his mind again on May 16, Pemberton finally decided to obey Johnston's orders and attempted to link up with him at Brownsville by countermarching his field force back through Edwards Depot. By now however, it was too late. Grant was already in contact with his lead division near Champion Hill.

Grant had rapidly moved his forces to the west toward Vicksburg, having left Sherman with two divisions in Jackson to complete the destruction of railroad tracks and stores. Grant therefore had about 29,000 men to oppose almost 23,000 Pemberton had brought out of Vicksburg. In a desperate battle with key positions changing hands several times, the Confederates were finally forced to retreat towards Vicksburg. Loring's division was cut off and would eventually link up with Johnston's army several days later. Pemberton tried to make a final stand at the Big Black River, but his forces were again routed and Grant's troops drove the Confederates into the Vicksburg defenses. Grant reestablished contact with the Federal fleet on the Yazoo River on May 18.

Grant ordered assaults to take the city on May 18, and again on May 22. The Confederate defenders proved obstinate however, and after suffering heavy losses, Grant settled down to a siege to starve the garrison into submission. In the meantime, Johnston was rapidly accumulating additional troops to attempt to relieve the siege. At one point his own and Pemberton's forces would actually outnumber the besieging Federal troops. However, coordinating an assault by Johnston with an attempted breakout by Pemberton proved impossible, and on June 15 Johnston notified the Confederate authorities that he considered saving Vicksburg hopeless.

By the end of June, the constant pressure and lack of food was beginning to tell on the Vicksburg defenders. On July 4, Pemberton formally surrendered his army of 2,166 officers and 27,230 enlisted men, 172 cannon and 60,000 small arms. Five days later another 7,000 Confederate troops would surrender to Major General Nathaniel Banks at Port Hudson, Louisiana.

The Mississippi River was once again open to sea and the Confederacy was effectively split in half. Grant's persistence and bold strategy had finally paid dividends.

Chapter Nine

The Eastern Theater: The Gettysburg Campaign

At the same time Pemberton was considering surrender terms, a far superior Confederate general was also experiencing defeat. General Robert E. Lee had just failed to break the center of the Federal line with Pickett's charge at Gettysburg.

Although the location and timing of the Battle of Gettysburg was almost an accident, the fact that a huge battle took place in Pennsylvania in July 1863 was certainly no accident. General Robert E. Lee, the commander of the Confederate Army of Northern Virginia, had chosen a course that meant a major battle was almost unavoidable.

Two months earlier, General Lee found himself in a dilemma. He had recently defeated the Federal Army of the Potomac in a daring example of superior generalship at the Battle of Chancellorsville. However, although Chancellorsville had been a decisive victory for the Confederacy, Lee himself knew that all he had really accomplished was bought a little more time. The battle had succeeded in repelling the Federals from much of Virginia, containing the all-important capital of the Confederacy at Richmond. However, Lee knew it would only be a matter of time before the powerful Federal host moved south again.

Lee was really only left with two viable choices, either to dig in and prepare to fight another defensive battle for Richmond, or to assume the initiative and attack. The second option was obviously dangerous, since the Confederates were significantly outnumbered, but it did have certain advantages. And Robert E. Lee was fast becoming known for his ability to pull off a victory against long odds.

Besides the advantage of retaining the initiative, the Confederate army was short of supplies, including food, clothing, and — shoes. A thrust into Pennsylvania would provide an opportunity to correct those deficiencies. An additional potential benefit was the possibility of gaining foreign recognition for the Confederacy, and strengthening the Northern Democrats, who were in favor of making peace with the South. Probably most decisive in Lee's mind however, was that his instinct told him to attack. Relinquishing the military initiative and enduring a siege were simply not his style.

Despite some objections from within the Confederate Cabinet, and misgivings on the part of the Confederate President, Jefferson Davis, Lee eventually was granted permission to undertake his Northern invasion. His first task was that of army reorganization. At Chancellorsville, Lee had lost his "right arm", General Stonewall Jackson, due to an accidental shooting by his own troops. Jackson seemed to have the uncanny knack of almost being able to read Lee's mind, and Lee was careful not to issue Jackson overly specific orders in executing his battle plans. But Lee had other very good commanders. Prior to the Gettysburg Campaign, the Army of Northern Virginia was divided into two large corps, or "wings," one under the command of Jackson, and the second by Lieutenant General James Longstreet, to whom Lee referred as his "Old Warhorse."

The reorganization divided the army into three corps, keeping Longstreet in command of the First Corps, and promotions of

Lieutenant Generals Richard S. Ewell and Ambrose Powell Hill to the command of the Second and Third Corps, respectively. Though both these commanders had been successful, and would go on to further successes, neither had quite that "gift" of Jackson's.

Lee also had an outstanding cavalry commander, James Ewell Brown (JEB) Stuart, who had literally "run rings" around his Federal counterparts. Now however, Stuart's highly inflated ego may have become responsible for his letting Lee down very badly in the upcoming campaign. Despite the shortcomings of the command structure of the Army of Northern Virginia, there can be no doubt that it was one of the most able and least afflicted by personal differences in the history of warfare. The same cannot be said for their Federal counterpart, the powerful but much misused, Army of the Potomac.

Major General George Gordon Meade's problems were considerable. The Army of the Potomac's latest commander, appointed only June 28, Meade would not have time to cope with his new responsibilities, yet this decisive struggle had been thrust upon him without warning. His army was large, well-clothed and well-armed, but was a "long-suffering" organization. In two years these Northern veterans had fought six battles, incurring five defeats and a draw that could only be called a bitter failure. During this period, they had served under as many commanders, each which had demonstrated command inferiority to his Southern counterpart. Yet in spite of these circumstances, the men in the ranks displayed no air of being a defeated army. Somehow they had forged a pride and cohesion which had been a trademark of the Southern forces from the beginning. The credit for this can be attributed to the army's corps and division commanders — men like Sedgwick, Sickles, Hancock, Reynolds, and Slocum. These subordinates would prove critical in the upcoming campaign.

The invasion of the North proper began on June 15, with the leading divisions of Ewell's Corps crossing the Potomac River near Shepherdstown, and entering Maryland. Officially still loyal to the Union, the state had provided several excellent combat units to Lee's army. On June 19, the Maryland-Pennsylvania border was crossed,

39

threatening the cities of Baltimore and Washington and ensuring that Meade would have to respond.

Lee apparently had no clear idea of where exactly he wished to fight the decisive battle. Orders to Ewell were to advance on a broad front, allowing progress and directions to be determined by the "development of circumstances."

Early in the Gettysburg Campaign, Ewell had executed a well-organized capture of three Federal garrisons near Winchester, Virginia. Thus, as a new corps commander, he had gotten off to an impressive start. Less impressive was the role played by Stuart's cavalry. Stuart had been rocked by a surprise advance of the Federal cavalry resulting in the Battle of Brandy Station on June 9, the largest cavalry fight of the war. Although the Northern horsemen had ultimately been driven back, Stuart's reputation had been tarnished. He may have been driven to perform another "glory ride" to redeem his reputation, and while off on this enterprise, left Lee's army virtually blind to Federal positions and movements.

Gettysburg was a small, prosperous farming town. When a Southern brigade first passed through it, its commander noted that it contained a shoe factory. On June 30, a Confederate brigade had been sent to appropriate some of these shoes, but had withdrawn because of a large Federal force seen heading towards the town. The corps commander, Lieutenant General Hill, did not believe the Federals could be so near, and raised no objections when Major General Henry Heth, leading one of Hill's divisions, asked if he could use his superior force to collect some shoes. This was the spark that set off the battle.

The following day, July 1, 1863, saw the first fighting at Gettysburg. The first encounter was between Heth's brigades and two dismounted Federal cavalry brigades under the command of John Buford. The meeting engagement quickly escalated at a rate that was beyond the comprehension of the commanders on the scene.

The opening fight, which started around 8 a.m., was at its heaviest in the ridges to the northwest of Gettysburg, first on Herr Ridge, then in and around McPherson Woods. The Union cavalry commander,

Brigadier General Buford, had correctly interpreted the Confederate intentions and had placed his dismounted breechloader-carrying troopers to delay the initial Southern advance.

At this time, the rest of the Army of the Potomac was spread out, heading for a defensive line along a river known as Pipe Creek. On the Confederate side, Ewell's Corps was still well to the north of Gettysburg, and Longstreet's was a day's march to the west. Stuart was still nowhere to be seen.

The Northern cavalry was quickly outnumbered and Buford urgently sent for help from the Federal commander of the left wing, Major General John Reynolds. When Reynolds arrived, he could see the dismounted cavalrymen being pushed back. He sent an urgent message to General Meade, informing him that the enemy was advancing in strong force on Gettysburg.

General Meade, still uncertain of the location of Lee's main force or the defensive position he should occupy, was reassured at Reynolds' confidence in making a stand. The brigades of Reynolds' Corps were rapidly becoming engaged in the McPherson Woods. It was only when the Confederates recognized some of the veteran regiments of his corps that they realized for sure that they had run into the Army of the Potomac and were not engaged with Pennsylvania militia.

Heth reported to Hill that he had encountered a strong Federal force. Hill, apparently ill but not incapacitated, sent back word that Pender's division was being sent to his support. The momentum of the meeting engagement was accelerating.

Around noon, the Federal forces controlled McPherson Woods, Seminary Ridge, the southern end of Oak Ridge, Barlow Knoll, and the town of Gettysburg itself. Shortly thereafter, Major General Reynolds, the Federal I Corps commander, was killed. Major General Howard, commanding the XI Corps, arrived ahead of his main body and assumed command. Howard's lead units arrived just in time, reinforcing Gettysburg to stall a major Confederate drive on the town.

Luckily, Hill had informed Lieutenant General Ewell, commanding the

Army of Northern Virginia's Second Corps, that Heth was heavily engaged at Gettysburg. Ewell responded by sending two divisions toward the town. The lead division of Major General Rodes hit Howard's men just as they had established their defensive line. Rodes' initial attacks were thrown back by coordinated defensive fire. The fight along Oak Ridge see-sawed back and forth as both sides were alternately repulsed by massed musketry or short range artillery fire.

About 2:30 in the afternoon, General Lee arrived at the summit of Herr Ridge to observe the fighting. He quickly canceled his initial orders to hold off any major Confederate attack until the army was fully concentrated. He saw Jubal Early's division come into view and concluded that much of the Federal position was vulnerable to becoming trapped between the divisions of Heth, with Pender in support, and the newly arriving division of Early. Being hit in front and flank did have a crushing effect on the defenders of XI Corps. They were routed from their positions and went streaming from Oak Ridge and Barlow Knoll back into Gettysburg. Quickly following, Pender's division pushed the I Corps troops defending Seminary Ridge back into Gettysburg and onto Cemetery Ridge, all in considerable confusion.

Fortunately for the Federals, Major General Howard had the foresight to leave one of his divisions on Cemetery Hill to construct defenses and secondly, Meade had instructed the highly capable and inspiring Major General Hancock to assume command. Hancock arrived on the field to observe the Federal forces streaming out of Gettysburg and onto Cemetery Hill. General Howard was desperately trying to stem the flow of the Union retreat. Hancock quickly decided to form a defensive line based on Cemetery Hill, but with additional divisions deployed on Culps Hill on the right, and a left flank extending down Cemetery Ridge as far as the Round Tops.

General Lee was eager to take advantage of the confusion evident in the Union retreat and attack in force. General Hill successfully argued however, that his own divisions had been too bloodied and were now too fought out to continue that day. Longstreet had arrived on the scene, but his own corps was still some distance away, so a continuation of the attack could only be carried out by Ewell's Corps. Ewell, who was used to receiving unmistakable orders from Stonewall Jackson, was not yet acquainted with Lee's style of issuing orders. Therefore, the discretionary phrasing of Lee's orders, i.e., "Attack if practicable," confused him. By the time Ewell finally realized that he had been ordered to take Culps Hill, the opportunity had already gone; the Federal divisions were already dug-in.

Meade, at his headquarters in Taneytown, was still trying to decide whether to pull back to Pipe Creek or not. However, General Hancock arrived and persuaded him that Gettysburg was the place to make the fight. Now convinced, Meade ordered all remaining corps of the Army of the Potomac to concentrate there and himself started out for Cemetery Ridge.

The fighting around the town finally died down the evening of July 1. With Lee and Meade now both on the field by early the following morning, the stage had been set, by a completely uncoordinated and uncontrolled sequence of events, for the biggest battle yet of the Civil War.

Lee's plans for the second day were for Longstreet to launch an attack on the Federal left, around the Peach Orchard and the Wheatfield. The attacks were to fall "en echelon" from right to left, with Ewell to coordinate an attack on Culps Hill if the opportunity arose.

Having had his own suggestion to Lee concerning a flanking movement to the Confederate right rejected, Longstreet encountered delay upon delay in moving his divisions into place and, when they were finally in the position Lee had wanted them, the situation had changed. The once weak Federal line along the Emmitsburg Road had been reinforced to the point that the Confederate divisional commanders thought it should not be attacked, but instead should move further to the right and attack Little Round Top. In hindsight, the failure of Lee or Longstreet to order an attack on Little Round Top was probably the greatest tactical error of the second day. Its commanding height would have allowed Confederate artillery to enfilade almost the entire Union line.

When the attack finally did get underway, a Confederate regiment under Colonel Oates did manage to capture the top of Big Round Top. Oates was convinced that the Round Tops held the key to the battlefield. However, the commanders on both sides were occupied with other concerns at the time. The Federals held onto the summit of Little Round Top in one of the better documented small unit actions of the war as the 20th Maine Infantry Regiment and 15th Alabama Infantry Regiment slugged it out on the wooded slopes. The Union left flank remained securely anchored.

The main Southern assault against the Federal left first pushed the Union troops out of the Peach Orchard, and then the Wheatfield and a rocky hill known as the Devil's Den. Furious fighting characterized this assault, as the soldiers on both sides seemed to sense that this was the decisive encounter. Fortunately for the Union, Meade's line allowed rapid reinforcement of threatened sectors because he had the advantage of interior movement. Thus, Little Round Top was quickly reinforced by V Corps, and VI Corps moved to seal the gap between the Round Tops and the more elevated sections of Cemetery Ridge to the north. After the second day's fighting died down in the evening, the Federals strengthened their prepared defensive positions along Cemetery Ridge and the Round Tops.

The terrible slaughter of the second day, as massive as it was, did not match what was to happen on the third. General Lee, desperate for a decisive victory, made a particularly bad decision. Some historians have suggested that Lee was suffering from the beginnings of a heart ailment that would eventually kill him, and thus clouded his military judgment that day. Whatever the reason, now Lee, having failed on both Union flanks, would decide to attack the center of the Federal line. What made matters worse was that Meade anticipated that decision on Lee's part. The attack would focus on a copse of trees, near the point now known as The Angle, in the center of Cemetery Ridge.

The attack would be preceded by a massive artillery bombardment of Cemetery Ridge, by a Grand Battery of 170 guns massed from the Peach Orchard to Seminary Ridge. Pickett's division would then spearhead the main assault of 15,000 Southern infantry, to storm and capture the Union center. A diversionary attack on Culps Hill was to be made to prevent Meade from shifting troops to reinforce his center. Longstreet made one last entreaty to General Lee to call off the attack, but failed. He knew the attack, forever known to history as Pickett's Charge, would fail disastrously.

The initial diversionary attacks on the Culps Hill sector began around 5 a.m. and subsided about 11 a.m., nothing of significance being achieved. The next two hours enjoyed a lull in the fighting, until at 1 p.m. the Confederate artillery opened. The 170 guns produced a noise that was absolutely deafening, but many of the Southern shots went long, falling into and creating more damage to units being held in reserve behind the lines than to the front line troops.

Although some Federal guns returned counterbattery fire, most conserved their ammunition for the imminent infantry attack. As more Union guns fell silent to conserve ammunition, the impression given to the Confederate artillery commander, Colonel Porter Alexander, was that his own fire was responsible for taking them out of action. It was during this lull that he advised Pickett that if he was going to advance, do it now because the situation would not improve any further. When Pickett asked Longstreet for permission to advance, all he could do was nod, probably overcome with emotion from knowing what would happen. Soon after the Federal guns fell silent, more than 12,000 veteran Southern infantry moved to the attack.

About 5,700 Federals were posted in defense of the half-mile front that was the focus of the attack of the divisions of Pickett, Pettigrew and Trimble. With parade ground precision, the Southern ranks moved forward, calmly filling the gaps being blown in the lines by Union artillery and continuing to advance.

The results are best told by the stark statistics of the charge. Of the 12,000 men who went forward, less than 5,000 would return. Pickett never forgave Lee for what happened. To Lee's credit, he shouldered the entire responsibility for the failure of the assault.

The bloodshed had not quite run its course, however. A pointless cavalry charge, ordered by Union Brigadier General Judson Kilpatrick, on the Confederate right flank was met with a bloody repulse, and so ended the third day's fighting. Losses during the three-day battle were staggering for both armies. Confederate losses were about 28,000 compared to 23,000 casualties for the Federals.

Expecting Meade to launch a counterattack, Lee ordered his army into a defensive posture. When July 4 came and went without a Union attack, Lee decided to retreat back into Virginia. The Battle of Gettysburg was over. Meade, suspecting Lee of trying to set a trap for him, was slow to pursue.

By July 14, despite some clashes with the Federal cavalry in rear-guard actions, Lee was back across the rain-swollen Potomac River. Meade crossed in pursuit but Lee managed to cross both the Rappahannock and the Rapidan rivers, and by August 4, the two armies were more or less back where they had started at the beginning of the campaign.

Chapter Ten

The Western Theater: The Tullahoma Campaign

For six months following the battle of Stones River, Bragg and Rosecrans had uneasily faced each other in Middle Tennessee. Despite repeated urgings from Washington, Rosecrans refused to budge from Murfreesboro until he felt his army was ready. When he did finally decide to advance on June 24, he did so with speed and skill.

Sending forces under Crittenden and Granger on diversionary movements to the east and west, Rosecrans sent his main force straight ahead toward Manchester, Tennessee. He was advancing in rough country with several easily defended passes to overcome, but a swift-moving advance by Colonel John Wilder's mounted infantry brigade, armed with the Spencer rapid fire, seven-shot repeating carbines, broke through Confederate forces at Hoover's Gap. Rosecrans was now squarely on Hardee's corps flank with a road open to his rear. Bragg had no choice but to fall back on his supply base at

Tullahoma and made preparations to defend against an attack by the Federal forces.

Having reached Manchester on June 27 however, Rosecrans again deceived Bragg by moving southeast instead of southwest and moved around Bragg's right flank. This movement now threatened the railroad that was Bragg's line of supply. With Granger and McCook in Shelbyville, directly north of Bragg's position, he was placed in a difficult position. After another raid on his railroad lines by Wilder's "Lightning" brigade, Bragg decided to yield Middle Tennessee to Rosecrans and retreated again across the Tennessee River.

Thus, with remarkably few casualties, Rosecrans had allowed Federal occupation of all of Middle Tennessee and taken more than 1600 prisoners. By July 7 however, the Washington authorities, elated with the dual successes at Gettysburg and Vicksburg, were already urging Rosecrans to advance again.

He would be ready to move forward again on August 16 in cooperation with a movement against Knoxville by Major General Burnside who had set out on August 15. Burnside's opponent, Major General Simon Bolivar Buckner, pulled out of Knoxville and Burnside entered unopposed on September 3. The loss of Knoxville cut the only direct Confederate rail link between Richmond and Chattanooga. Burnside then moved against the Southern forces guarding the Cumberland Gap, and took 2,500 prisoners. He then decided that he no longer needed to support Rosecrans as originally ordered since he had learned that Bragg was in full retreat.

As it turned out, Rosecrans would have been grateful for Burnside's support. The most obvious route to his continuing advance on Chattanooga was to the north, but Bragg had it well-defended. To deceive Bragg that the northern route was the one he intended to use, Rosecrans sent three brigades with orders to create the impression that a large force was preparing to cross the river at that point. Bragg moved reinforcements to cover the anticipated movement. Instead, Rosecrans crossed the Tennessee River at Bridgeport, about fifty miles south, virtually unopposed.

For speed of movement and maneuvering options, Rosecrans then split his army into three columns. Crittenden's corps was sent directly north to Chattanooga. McCook's corps was ordered to take a southern detour through Winston Gap. Thomas's corps, which Rosecrans accompanied, moved straight through the middle. Once more, Bragg had been completely outmaneuvered, and was forced to quickly evacuate Chattanooga. Crittenden entered Chattanooga without a fight.

It is at this point that Rosecrans became somewhat overconfident. Bragg sent out fake "deserters" who spread the story that the Confederate army was completely demoralized and in full retreat. Rosecrans optimistically believed these stories to be true. In fact, Bragg was looking for Rosecrans to overextend himself so he would open himself up for a counterpunch.

With the Federal army split into three widely separated columns, it appeared the opportunity might present itself to defeat each one in detail. It was a good plan, but delays and disorganization within Bragg's own army would prevent it from being successfully executed. Opportunities to crush first Thomas and then Crittenden, were wasted due to misunderstood orders or simply disobeyed by untrusting subordinates in Bragg's army.

When Rosecrans finally realized the danger his own army was in, he desperately tried to reunite the widely separated columns. By September 18, he had been successful in concentrating most of his forces east of the ridge near the Rossville Gap, about seven or eight miles east of Chattanooga on the banks of Chickamauga Creek. Here one of the bloodiest battles of the Civil War would be fought; a battle that would give the Confederates a much-needed victory following the triple reverses of Vicksburg, Gettysburg, and Middle Tennessee.

Chapter Eleven

The Western Theater: Chickamauga and Chattanooga

Bragg's Army of Tennessee had received reinforcements from Buckner's Knoxville command and two divisions of Longstreet's corps sent by rail from Lee's Army of Northern Virginia would be arriving to bolster Confederate troop strength to about 65,000 men. Rosecrans had approximately the same number under his command. Thus Chickamauga would be one of the few large battles of the war fought with approximately equal numbers on both sides.

The heavily wooded battlefield left little room for maneuvering and the two-day battle seemed as if it would degenerate into a simple slugging match. On September 20, confusion in orders left a gaping hole in Rosecrans' right flank which Longstreet exploited with an assault by four divisions. Almost half of the Federal army was routed and hastily retreated towards Chattanooga along with Rosecrans and most of his staff. A total Union disaster was averted by the stand of Thomas' corps on Snodgrass Hill which held the left wing together long enough to organize an orderly retreat. This earned Thomas the nickname of "The Rock of Chickamauga."

It was a great Confederate victory, but a very costly one. Although Federal losses had exceeded 16,000, the Southern army had lost more than 18,000 casualties. Bragg felt his troops were in no condition to implement a rapid pursuit of the retreating Federals, who were allowed to fall back into defensive positions in Chattanooga. Occupying the heights overlooking the city, Bragg confidently waited for the Federals to either leave...or starve.

The situation did look bad for the Union forces. Since the Confederate Army of Tennessee controlled the heights of Missionary Ridge and Lookout Mountain, the only Federal supply line was a very tenuous wagon route through the mountains, totally inadequate to supply a large army.

Bragg was not without his own problems, however. His own supply situation, while not threatened by Federal troops, was still a difficult one. It might be debated as to whether the Southern troops or Northern troops were hungrier. Additionally, Bragg was engaged in an on-going fight with his own subordinates. Especially since Chickamauga, many of his generals were openly expressing displeasure with his conduct of the battle and the way he was running things now. Bragg found a scapegoat in Lieutenant General Polk, and relieved him of command. Another corps commander who came from Lee's Army of Northern Virginia, D.H. Hill, was also removed by President Davis because of disagreements with Bragg. Major General Nathan Bedford Forrest, whose dismounted cavalry had served so well on Bragg's right flank at Chickamauga was so disgusted with Bragg and his treatment that he threatened to kill him if he ever again crossed his path. Davis gave Forrest an independent command and removed him from Bragg's authority. Bragg, apparently still having Davis' confidence, conducted a major reorganization of his army; trying to break up concentrations of "anti-Bragg" elements by placing "pro-Bragg" supporters in command where possible.

The Federals were also looking for their own "sacrificial lambs" for the

Chickamauga disaster. McCook and Crittenden, two of Rosecrans' corps commanders, were relieved of command for fleeing the field (although Rosecrans had, by some accounts, beaten them both back to Chattanooga). Lincoln realized however, that despite the loss at Chickamauga, the Federals still controlled Chattanooga and must continue to hold it. Major General Hooker and 20,000 men would be detached from the Army of the Potomac to Chattanooga and Sherman would bring another five divisions from the west as reinforcements. Soon, the besieged Federal army inside Chattanooga would outnumber Bragg's besieging forces.

For a month after Chickamauga, Rosecrans would remain in command. However, this was only a temporary arrangement. Lincoln had to determine the best way to put Major General George Thomas, the only Federal corps commander to escape any blame for Chickamauga and Grant who was now relatively idle following Vicksburg, to best use. The solution to this problem lay in the creation of the Military Division of the Mississippi, consisting of the Departments of the Cumberland, Ohio, and Tennessee, and placing Grant in charge. Thomas was given Rosecrans' former command, the Army of the Cumberland.

When Grant arrived in Chattanooga on October 23, he found the Federal army on the verge of starvation. The men were surviving on quarter rations and most of the animals were already dead. However Thomas and his chief engineer, "Baldy" Smith had been devising a plan to reopen their supply line. A small fleet of flat assault boats had been constructed. Their purpose was to float downstream to Brown's Ferry and take control of that point by surprise. The boats would then be converted into a pontoon bridge and additional troops would cross, take Raccoon Mountain to the west, and secure a new Federal supply line.

On October 27, during the early morning hours, 1500 men floated silently downstream on the sixty wooden boats and captured the unaware Confederate pickets. Hooker arrived from Bridgeport with another two divisions to secure the bridgehead, and the famous "Cracker Line" was opened. 40,000 rations were delivered to the

Federal troops on October 30 — one week after Grant's arrival. As the Union army was resupplied, it became clear that Bragg's strategy of starving out the Federals had failed and his options were quickly running out.

By the end of November, Bragg had detached Longstreet (another corps commander he could not get along with), to try and retake Knoxville from Burnside. That was when Grant decided to take action. He now had Hooker's and Sherman's troops available, as well as Thomas' Army of the Cumberland. He would attack Bragg's position on Missionary Ridge. Sherman would move around to hit Bragg's right flank and Hooker would attack his left. It was intended that Thomas would simply demonstrate against the center of Bragg's line to prevent reinforcements from being sent to either of the flanks.

Sherman's three divisions however, were brought to a standstill by Cleburne's division. Hooker had made progress on the left, but it was Thomas' soldiers, despite orders to the contrary, that attacked directly up the front of the ridge and split Bragg's army in half. In an attack against what should have been an impregnable position, Bragg's demoralized troops were sent streaming south into Georgia.

Longstreet was faring no better at Knoxville. The Federals had prepared extremely strong entrenched positions and the Confederates had been unsuccessful in detecting any weaknesses. However on November 27, after learning of Bragg's defeat at Chattanooga, Longstreet decided that he must launch an assault for the dual purpose of helping draw Federal troops away from pursuing Bragg and to help make his own withdrawal easier when that necessity came.

On the morning of November 29, the assault went forward. However, poor reconnaissance and planning resulted in the Confederates being trapped in a nine-foot ditch (which staff officers had reported as being only five-feet deep), with no scaling ladders. The attempt to press forward was hopeless. Some attempted to climb up on others' shoulders to fight their way out but the men were raked with artillery and were helpless targets for the Northern defenders. When

Longstreet called off the attack, he had lost 813 casualties to a total of only 13 for the Federals.

Learning a few days later that Sherman was headed his way with six divisions, Longstreet decided to call it quits and took his command back to Virginia to rejoin Lee. Davis meanwhile, had no choice but to relieve the discredited Bragg and placed the long suffering Army of Tennessee temporarily in the command of Lieutenant General Hardee in its winter camp at Dalton, Georgia.

Chapter Twelve

The Trans-Mississippi: The Red River Campaign

The fall of Port Hudson in July 1863 eliminated Confederate control of the lower Mississippi River and freed the forces of the newly created Department of the Gulf for employment elsewhere. Major General Nathaniel Banks, the commander of this department, was in agreement with Grant and Admiral Farragut that an expedition against Mobile, Alabama would be the most effective means of rendering support to the proposed operations against Bragg at Chattanooga, which was the highest priority of the Federal forces in the Western Theater.

The authorities in Washington however, had a somewhat different view of the situation. General Halleck, probably at Lincoln's insistence, directed Banks to move his forces against the Confederates in Texas. There were undoubtedly more political considerations than military in this redirection. It was felt to be very important that a Federal presence be re-established in Confederate Texas.

A Federal attempt to move against Sabine Pass at the mouth of the Sabine River in September 1863 ended in dismal failure. A second attempt to reach the Sabine River by an overland march was terminated because of anticipated supply difficulties in hostile territory. Banks fell back upon naval operations to reduce the Texas Gulf Coast ports. Between November and December 1863, Brownsville, Corpus Christi, and Fort Esperanza were occupied by Federal forces. By the beginning of 1864, the only major port in Texas still in Confederate hands was Galveston. In early January, Banks began operations against it also.

Hardly had the Galveston operations begun than Halleck directed Banks to resume the delayed Red River operation against Shreveport, Louisiana. This expedition was to be of a larger scale than the earlier ones, with Banks to receive support from Steele's Federal forces in Arkansas, detachments of Sherman's command in Mississippi, and gunboat support from Farragut under the command of Admiral Porter.

Except for a few weeks in late March and April, when spring rains swell its depth, the Red River was generally non-navigable above Alexandria, Louisiana. Bearing this in mind, Banks planned the operation to get underway by mid-March. However, the Federal plan suffered from serious defects. 10,000 troops from Sherman, to be transported by Porter's gunboats, were to rendezvous with Banks' 17,000 at Alexandria, some 100 miles deep into Confederate controlled territory. Even worse, the junction with Steele's 15,000 troops from Arkansas was to occur at Shreveport, another 150 miles behind Confederate lines. Although Banks was not happy with the situation, and unable to convince his superiors of the pitfalls that might lie ahead, he resolved to try and carry out his orders to the best of his ability.

The overall Confederate commander for the Trans-Mississippi Theater was General Edmund Kirby Smith. Smith could muster about 25,000 men to oppose the Federal operations. Major General Richard Taylor, son of former President Zachary Taylor, was placed in command of Confederate field forces.

Sherman's detachment, under the command of Major General A. J. Smith, were the first to arrive via Porter's gunboats at Alexandria, Louisiana on March 18. Banks' command arrived a week later. On March 27, Banks received new orders from General Grant. Grant required that the operations against Shreveport must be concluded by April 25 because he would need all the troops for operations against Atlanta and Mobile by early May. Based on this new development, Banks considered calling off the campaign. However, he was optimistic that the Confederates would not be able to concentrate in time and even if they were able to do so, that they might choose not to contest the Federal occupation of Shreveport.

By April 3, the Red River had risen enough to allow Banks' transports and thirteen of the smaller gunboats to pass the rapids above Alexandria. The Confederates were still not fully concentrated and were gathered at a plantation about forty miles northwest of the city. Taylor awaited the arrival of two divisions from Sterling Price's forces, under the command of Brigadier General Thomas Churchill before moving against the Federals.

Leaving about 5,000 men to provide security for his line of communications, Banks set off on April 6 with a force of about 24,000 men toward Mansfield. On April 8, his advance elements encountered Taylor's army of about 16,000 occupying an advantageous position on the edge of a small clearing about two miles south of Mansfield. The battle did not begin in earnest until about 4:00 p.m.

The Federal troops were sent forward to assault the Confederate positions without proper support for their flanks. This error, in combination with some poor tactical decisions on the part of the Federal commanders, allowed the Southern forces to put the Federals to flight after two hours of bitter fighting. Under the counterattack of Taylor's best troops, the Louisiana division of Alfred Mouton, the Federal divisions of Landram and Cameron fell apart. Only the timely arrival of Emory's division saved the Federal forces from complete disaster.

Banks withdrew during the night to Pleasant Hill, about nine miles southwest of the Confederates. Emory's division was deployed to the front while A. J. Smith's fresh troops formed a second line and reserve. The troops of Landram and Cameron, who had taken such a beating, were sent to the rear to guard the wagons and would take no part in the upcoming fight.

Taylor wanted desperately to complete the destruction of the Federal army. He put his tired troops on the road in pursuit and by 1:00 p.m. on April 9 had reached the vicinity of Pleasant Hill. The Southern forces were allowed to rest for about two hours while Taylor devised a plan of attack. Taylor deployed Churchill's two divisions, unengaged the previous day but still very tired from two day's hard marching, on his right flank with orders to assault the Federal left. Walker's and Mouton's divisions were deployed in the center intending to pin the Union defenders in place. Meanwhile, the Confederate cavalry was to move around the Federal right and place themselves in a position to cut off Banks' expected retreat route.

Unfortunately for Taylor, Churchill's line of assault did not go deep enough against the Federal flank, leaving his own flank exposed to a counterattack by the Union reserve. The Confederate assault was repulsed with heavy loss.

The following morning found Banks again having retreated. Kirby Smith reached the battlefield and left Taylor with Mouton's division and the cavalry (about 5,200 troops) to continue to harass Banks withdrawal while taking the remainder to move against Steele in Arkansas. Banks, by this time, had abandoned his attempt to capture Shreveport since he could now expect no help from Steele and he must soon return Sherman's troops.

After several skirmishes and minor engagements between Banks' forces, Porter's fleet and Taylor's pursuing troops, the Federals finally escaped back down the Red River. Sherman's troops were embarked for Vicksburg on May 21-22, and on May 26, the remainder of Banks' command reached Donaldsonville, Louisiana.

The Red River Campaign had been a complete failure for the Federals. Banks was relieved of command and Kirby Smith fired Taylor as a result of some angry correspondence between the two Confederate commanders.

Chapter Thirteen

The Eastern Theater: Lee and Meade

Ever since the retreat from Pennsylvania, Lee had looked for an opportunity to resume the offensive against the Federal Army of the Potomac. However, detaching most of Longstreet's corps to Bragg had reduced Lee's strength to less than 50,000 men, only about half as many as Meade, and made taking the offensive impossible. Upon learning that Meade had detached two corps to aid Grant at Chattanooga however, Lee decided to try and repeat his success of a year ago against Pope, who had been in a similar position at that time.

Lee began his march north on October 9, 1863. As expected, Meade like Pope, fell back from the constricting "V" between the Rapidan and Rappahannock rivers. Unlike Pope however, Meade did not stop to contest Lee's crossing of the Rappahannock but continued to fall back along the Orange and Alexandria Railroad.

On October 14, Lieutenant General A.P. Hill saw what he thought was an opportunity to bag half the Union III Corps, caught mid-way across Broad Run and apparently milling in confusion. Without conducting a proper reconnaissance, he rushed two of the brigades from his corps directly into a well-laid Federal trap.

Warren's II Corps was hidden behind a railway embankment and when the Confederate troops had advanced far enough, they fired directly into their flank. Not realizing the strength of the Union position, the Southerners charged the Federals and were cut to pieces. It was two brigades against three divisions and the Confederates lost heavily. About 1400 were killed and wounded and another 450 taken prisoner. This action is generally known as the battle of Bristoe Station.

Lee continued his pursuit but discovered that Meade had entrenched in a very strong position along the Centerville-Chantilly ridge. With the approach of cold weather and an inadequate supply line, Lee decided he could not remain in the area and withdrew on October 17. Meade sent his cavalry in pursuit but the Federal troopers were ambushed by Jeb Stuart's Southern horsemen and driven back.

Meade moved forward again, but not very rapidly because he was forced to repair the railroad along the way. By the end of October however, he was back on the banks of the Rappahannock where Lee had stopped and entrenched. Although it was a strong position, an unexpected night attack on Kelly's Ford succeeded in turning Lee's position and forcing him back across the Rapidan.

By the end of November, Meade was receiving intelligence reports that Lee's army now numbered only about 40,000 troops (Lee actually had 48,000) against Meade's own 84,000 effectives. He decided to move against Lee and so crossed over the Rapidan into the fringe of the Wilderness where Hooker had come to grief seven months ago. Meade employed no diversionary movements but depended on speed and superior numbers and hoped to catch Lee unprepared.

There were the inevitable delays however, and by the time the Federal troops arrived at Mine Run, they found Lee's Army of Northern Virginia strongly entrenched behind seven miles of earthworks, all with cleared fields of fire and overlapping fire support. Still, if he could only locate a weak spot, Meade was determined to break Lee's lines. When reports came in that indicated weakness on Lee's flanks, Meade ordered a dawn assault by Warren's II Corps while Sedgwick's VI Corps conducted a diversionary attack with artillery on the opposite end of the line. When Warren reported that the attack no longer looked feasible, Meade went forward to see for himself. Agreeing with Warren's assessment, he canceled the attack.

Meade abandoned his positions after sunset on December 1. Lee had ordered an attack for the morning of December 2, but the Federals were already gone. Lee again began a pursuit, but Meade's head start allowed him to recross the Rapidan. Both sides then went into winter camps.

Chapter Fourteen

The Eastern Theater: Grant Takes Command

On March 9, 1864, President Lincoln promoted U.S. Grant to the newly revived rank of Lieutenant General and on March 12 was made General in Chief of the Armies of the United States, taking over the strategic direction of the Federal war effort. Major General Hallack was made Chief of Staff to oversee the administration and logistics details and so removed from Grant the burden of paperwork and allowed him to concentrate solely on overall Federal strategy.

Grant's plan was fairly simple. Sherman, who took Grant's old job in command of the Federal armies in the Western Theater, would advance against the Confederate Army of Tennessee defending Atlanta while Grant would do the same against Lee's Army of Northern Virginia and Richmond. Once the armies were engaged, there would be no letup in pressure against the remaining Confederate forces. Whoever reached their objective first would then join the other for final offensive operations.

Grant shook up the command of the Army of the Potomac, removing many "rear area" units from their cushy assignments and putting them into combat organizations. He also brought Phil Sheridan east to command his 13,000 Federal cavalry. Grant was ready, by the beginning of May, to move south. After giving Sherman his go-ahead, the Army of the Potomac was again ready to go into action against Lee and his Army of Northern Virginia.

Chapter Fifteen

The Eastern Theater: The Forty Days

On May 4, 1864, Grant began his advance by taking the same route followed by Hooker and Meade through the Wilderness around Chancellorsville. He hoped that by moving at a rapid pace, he would be out of the tangled undergrowth before Lee could react.

Major General Ben Butler was ordered to cooperate in this movement by advancing his Army of the James up the Yorktown peninsula to threaten Richmond from the south and east. This movement, Grant hoped, would provide a diversion to the main movement by the Army of the Potomac.

Things did not quite work out as planned for Grant. Lee reacted to Grant's movement and brought him to battle before he had cleared the Wilderness. This three day slug-fest was very confused, but the battle went heavily against the Federals.

On May 5, the V Corps, now commanded by Major General Warren, collided with Ewell and initially pushed him back. Ewell later counterattacked and Warren was almost outflanked by Hill who was eventually stopped by Hancock's II Corps. On May 6, Hancock pushed forward, but was stalled by Longstreet's arrival. A Confederate attack about mid-day punched into Hancock's left flank, but he eventually managed to stabilize his line. Burnside's IX Corps spent most of May 6 lost in the undergrowth, but eventually managed to make contact with Hill who repulsed his attack. Near sundown, a Confederate assault column led by Brigadier General J. B. Gordon smashed into and began rolling up the Federal right flank, but the attack fizzled out due to lack of support and the onset of darkness. May 7 was occupied with both sides digging-in.

In spite of the confused nature of the fighting, it was clear that Grant had been as decisively beaten as his predecessors. He had taken 17,666 casualties and had inflicted only about 7,800. Both Grant's flanks had been turned and Lee stood squarely in his front. If history was to repeat itself, Grant would retreat, a new general would be appointed, the army reorganized, and sooner or later the whole process would have to be repeated again.

But Grant was not a Pope or a Hooker. He pulled out of his lines, but instead of retreating, he moved down the Brock Road, toward Spotsylvania. Grant was the "killer arithmetician" Lincoln had been looking for. He knew the North could afford take the losses and replace them; the South could not.

Lee was not surprised at Grant's move, because if he could take and hold Spotsylvania, he would be between Lee and Richmond which would then force Lee to attack a numerically superior force in an entrenched position. The Confederates narrowly beat the Federals in the race to Spotsylvania and was able to beat back the Federal attacks on May 8.

The following day, May 9, Grant lost one of his best subordinates, Major General John Sedgwick, to a Confederate sniper's bullet. Ironically, Sedgwick had been trying to instill confidence in his troops with the words, "they couldn't hit an elephant at this distance," just seconds before he was struck down.

May 10th and 12th were days of heavy fighting, particularly around a salient in the Confederate lines known as the Mule Shoe. On May 12, Hancock launched a successful assault and penetration, capturing three generals, thirty artillery pieces and almost an entire division of troops. Fierce fighting allowed the Confederates to re-stabilize their lines at the base of the salient. Despite some additional fighting the following week, Grant was unable to find any other weaknesses in Lee's lines. On May 20 therefore, he began another flanking movement.

In the meantime, additional Federal thrusts were meeting with little success. Butler's Federal Army of the James had allowed itself to become bottled up in the Bermuda Peninsula and was taken out of the equation for Grant's continuing drive on Richmond. Sigel's Federal forces in the Shenandoah Valley had met a reverse at the hands of Major General Breckinridge at the battle of New Market on May 15. Breckinridge was ably assisted in the battle by the Corps of Cadets from the Virginia Military Institute. The only good news from a Federal perspective was that now Grant had an excuse to relieve Sigel and replace him with Major General David Hunter.

With Butler out of the picture, Grant's task was made even more difficult. However on May 20 he began another turning movement by sending Hancock's corps towards Hanover Junction. Grant had hoped that Lee would try and attack this isolated corps, bringing him out into the open where the remainder of the Army of the Potomac could defeat him outside his entrenchments. Lee would not take the bait however, and marched south so that when Grant arrived on the North Anna River, he found Lee already entrenched in another strong blocking position.

At this point, Lee set a clever trap which Grant was slow to recognize. The Confederate army was deployed in a "wedge" south of the North Anna River, with the apex of the wedge touching the river at Ox Ford. The Federal corps of Wright and Warren crossed upstream and

Hancock's corps, downstream of Ox Ford. While moving forward, they suddenly realized that Lee's position would enable him to fight a holding action on one side of this wedge while moving the bulk of his army to defeat part of the Army of the Potomac in detail. Almost in a state of panic, the Union army began entrenching at a frantic rate to protect itself against such an attack. As luck would have it however, the attack never came. Lee had purportedly been taken ill by an intestinal complaint and could not personally direct his forces. Apparently, his subordinates did not fully comprehend what he had planned to do and so no Confederate attack took place. Grant disengaged from the potential trap and moved once more to the left.

After a large cavalry clash, Grant moved sideways once again toward Cold Harbor, where he planned to link up with Smith's corps that had been detached from Butler's Army of the James. Lee again anticipated the move and ordered his cavalry to hold Cold Harbor until the infantry could be brought up. May 31 was an all-day cavalry fight with Sheridan finally succeeding in taking the position as night fell. Sheridan had noted Confederate infantry arriving and sent a message to the effect that he did not think he could hold. Grant and Meade sent back word to Sheridan to hold "at all hazards."

Lee, who wanted to recover the position and roll up the Federal left before Grant was in a position to do the same to his own right, ordered a Confederate attack. However, Lee was apparently still suffering from his illness and unable to properly oversee the operations. The assault was mismanaged and by mid-morning on June 1, Wright's corps had arrived to relieve Sheridan and the Federal position held. Lee then abandoned the idea of recapturing the position, but again entrenched in strong defensive works.

From June 1-3, Grant ordered a series of ill-advised frontal assaults against Lee's entrenchments and was repulsed each time. The main assault of June 3 was particularly bloody, with Grant losing about 7,000 men to the Confederates 1,500 in a very short time span of only a few minutes. Even though the Federals could more easily replace battle losses, Grant could see that these loss ratios were not in his favor.

The opposing lines stabilized for ten days at Cold Harbor while Grant considered his options. Grant's critics were quick to point out that Grant was now in the same position McClellan had been two years earlier. However, Grant had paid for that privilege with over 50,000 casualties or about 41% of his original strength.

Nevertheless, Grant's plan was still essentially working. He had kept Lee tied up with constant pressure and no Army of Northern Virginia troops were being sent to help Johnston in front of Atlanta where Sherman was still making steady progress. And, even though absolute Southern losses were much lower than those of the Federals, Lee had still lost about 27,000 men, or 40% of his own strength at the beginning of the campaign and several irreplaceable general officers had been killed, seriously wounded or captured. These were losses the Confederacy would not be able to replace.

On June 12, Grant did his last "side-slip" crossing the James River and attempting a new approach at Richmond from the south through Petersburg. Although delays and tactical failures on the part of his subordinates deprived Grant of the full fruits of this last maneuver, Lee was now bottled up inside his Richmond defenses. Grant undertook siege operations that would last for another ten months, but what Lee feared most had happened. He was now completely tied down defending Richmond and would no longer be able to conduct any major offensive operations against Grant.

Chapter Sixteen

The Eastern Theater: Early's Washington Raid

If Lee could only divert some of Grant's strength from his front, he might still be able to find a way to destroy him by offensive action. With this in mind, he detached Lieutenant General Early with four infantry divisions and a cavalry division to undertake an offensive in the Shenandoah Valley.

With this new Confederate "Army of the Valley," Early moved out on June 7, 1864. On June 18, David Hunter's Federal forces were defeated at Lynchburg, and on June 27, Early reorganized his forces for a thrust into the North at Staunton. He carried with him about 10,000 infantry and 4,000 cavalry.

He crossed the Potomac on July 5 into Maryland. Upon learning that the Confederates seemed to be making a serious movement, reinforcements from the Army of the Potomac were ordered by Grant to Baltimore and arrived on July 7. Rickett's division joined a scratch brigade of infantry and cavalry under Major General Lew Wallace near Frederick, Maryland. Here was fought the battle of Monocacy on July 9. Wallace was forced to fall back to Baltimore after delaying Early.

Early utilized some of his cavalry to protect his line of communications, and sent a cavalry brigade to threaten Baltimore. With the remainder of his forces, he marched on Washington, D.C. He reached the outskirts of the capital on July 11 about midday. He saw that its defenses had been reinforced but spent the rest of the day looking for a weak point to launch an assault to take place the following day.

That night Early learned that the Federal VI corps had arrived to strengthen the capital's defenses, and delayed his planned attack. After heavy skirmishing around Fort Stevens, Early concluded he lacked the necessary strength for a successful assault and withdrew that night.

The Federal pursuit was disorganized and Early took advantage of the situation to renew his offensive operations. He met and defeated Crook's forces at Kernstown on July 23-24 and moved two cavalry brigades to Chambersburg, Pennsylvania and burned the town in reprisal for Federal depredations on July 30.

Early's raid convinced Grant that he would have to take more drastic action to eliminate Confederate use of the Shenandoah Valley for strategic diversions. He made plans to put Phil Sheridan in command of newly reorganized and consolidated Federal forces in the area.

Chapter Seventeen

The Eastern Theater: Sheridan's Shenandoah Valley Campaign

As a result of the embarrassment of Early's Washington Raid, the Federal authorities set up the Middle Military Division and placed it under Major General Phil Sheridan who took command on August 7, 1864. Sheridan reorganized the various forces under his command to include a cavalry corps of three divisions. His effective strength was about 48,000 men.

Meanwhile, Early's Army of the Valley, with four infantry divisions and a division of cavalry, was to be reinforced by mid-August with Kershaw's infantry division and Fitz Lee's cavalry division. These reinforcements, under the command of Richard Anderson, were to support Early's operations east of the Blue Ridge. With these additional troops, Early's strength was about 23,000 infantry and cavalry, although the Federal estimates were that he might have as many as 40,000 men. Therefore, Sheridan was ordered to assume the defensive for the time being.

A time span of about five weeks were spent in maneuvering by both sides before Sheridan and Early finally met in battle. Because of lack of activity on Sheridan's part, Anderson's forces were ordered to return to Lee. However, Fitz Lee's cavalry remained in the Valley, leaving Early with about 12,000 infantry and 6,500 cavalry.

The return of Anderson to Lee was what Sheridan had been waiting for. This, coupled with overconfidence on the part of Early, led to his defeat at Winchester on September 19. Early fell back to Fishers Hill where Sheridan again defeated the Confederates on September 22.

The demoralized Southerners reached Browns Gap where they were again reinforced by Kershaw's division. Sheridan reorganized his cavalry with Brigadier General George Custer and Colonel William Powell as division commanders following the dismissal of Averell for lack of aggressiveness and Wilson's transfer to Sherman.

Sheridan believed he had ridded the Valley of any threat by Early and was in the process of transferring troops to reinforce Grant. These troops had to be recalled when Sheridan learned that Early was still in the area.

In a brilliant dawn surprise attack, Early's numerically inferior forces struck the Federals in camp at Cedar Creek on October 19. Despite initial success in driving the Federals from their positions, Sheridan arrived on the scene and helped rally his men to counterattack and rout Early's command. This was the last major action in the Valley.

Sheridan detached most of his infantry to rejoin Grant and Sherman, but kept his excellent cavalry corps in the divisions of Custer and Devin (about 10,000 troopers). Meanwhile, Early was left with only two brigades under Wharton (about 2,000 men) and two artillery battalions.

Sheridan decided to eliminate this remaining Confederate force. At Waynesboro on March 2, 1865, Custer's division overran and annihilated these remaining troops. Early managed to escape with several others, but his military career was ended.

Chapter Eighteen

The Western Theater: The Atlanta Campaign

At the same time Grant was beginning his drive against Lee, Sherman began moving forward against the Confederate Army of Tennessee, now under the command of General Joseph E. Johnston. Following the tenure of Braxton Bragg, Johnston exhibited real concern for his men and he soon became extremely popular among the troops. He gave the entire army furloughs, one-third at a time and announced amnesty for deserters. The troops were resupplied with rations, clothing, and new shoes. Discipline was maintained, and even more importantly, even-handedly administered. The army was restored to fighting trim.

The Army of Tennessee at the time comprised two infantry corps under the command of Lieutenant Generals William Hardee and John Bell Hood (a third corps under the command of Lieutenant General Leonidas Polk would soon join the army). The cavalry was commanded by Major General Joe Wheeler. Although Johnston's army was outnumbered almost two-to-one, he held excellent defensive terrain and was confident of disrupting Sherman's plans to take Atlanta.

Sherman's forces consisted of three Federal armies: the large Army of the Cumberland, under Major General George Thomas; the Army of the Tennessee, under Major General McPherson; and the much smaller, corps-sized Army of the Ohio under Major General Schofield. Later in the campaign, Blair's XVII Corps would join Sherman after Johnston had been forced to retreat beyond the Etowah River.

Throughout May and into mid-July, Johnston conducted a series of skillfully managed retreats while delaying Sherman's advance. He was also looking for Sherman to make a mistake that would allow him to move against a portion of his command and crush it before it could be reinforced.

Following the actions around Resaca in mid-May, Johnston thought he might have the opportunity he was looking for at Cassville on May 18. There, he had succeeded in deceiving Sherman as to his true line of retreat and prepared to fall upon the Army of the Ohio which was separated and subject to defeat in detail. However, Hood, becoming concerned on having Federal cavalry appear on his flank, called off the attack.

The next major encounter was at Kennesaw Mountain where Sherman, becoming impatient at his lack of progress in his flanking maneuvers, ordered a series of frontal assaults against Johnston's entrenched troops. Losing about 2,000 casualties to Johnston's 500, Sherman later justified the assault by asserting that it had shown Johnston that he would attack entrenchments and that Johnston must therefore keep them well-manned.

By mid-July, Sherman was across the Chattahoochee River and only six miles north of Atlanta. The Confederate President, Jefferson Davis, was unhappy with Johnston's strategy and on the eve of a major battle, replaced him with a more aggressive commander, John Bell Hood. When Sherman learned of the change, he was pleased with his new prospects.

Hood's first battle as commander of the Army of Tennessee was Peachtree Creek on July 20. In an attempt to hit the Federals while they were in the process of crossing Peachtree Creek, mistakes and misunderstandings on the part of Hood's subordinates prevented the Confederates from capitalizing on the situation. Here Hood lost about 4,800 men to the Federals' loss of 1,800.

On July 22, he saw another opportunity, this time to strike McPherson's left flank by sending Hardee's corps on a night march. Delays and bad management resulted in another 8,000 casualties for the Confederates to 3,700 Federals at the Battle of Atlanta (or Hood's Second Sortie).

On July 28, at the battle of Ezra Church Hood attacked again, this time losing about 5,000 men to only 600 Federal casualties. Hood was indeed showing aggressive behavior, but Sherman was winning all the battles. July ended with Hood finally deciding that attacking Sherman was not to his advantage, and withdrew behind Atlanta's fortifications.

Sherman however, did not intend to attack Hood behind his fortifications, but instead set about cutting his supply lines with the Federal cavalry. The cavalry failed miserably at their task, so Sherman decided to move from his base and destroy the railroads with his infantry.

When Hood was unable to prevent the inevitable at the battle of Jonesboro, he ordered the evacuation of Atlanta on September 1. Ordering the destruction of military stores that could not be saved, the city of Atlanta was wracked with a series of powerful explosions as the soldiers blew up locomotives, factories, and eighty-one railroad cars loaded with ammunition. The battered Army of Tennessee slipped away during the night, to regroup beyond contact with Sherman's Federals. The victorious Union army entered the city the following day.

In Washington, D.C., President Lincoln awaited word of Sherman. His political future and the future of the war had appeared uncertain. With Grant stalled by Lee before Richmond and Sherman's problematic drive on Atlanta, the strength of the Democratic peace party, led by

Lincoln's former general, George B. McClellan, had grown. Lincoln needed victories to ensure re-election and continuation of the war effort. They had not been forthcoming to this point in time. Finally a telegram from Sherman, given to Lincoln on the afternoon of September 2, 1864, changed all that. It began with the simple statement that, "Atlanta is ours, and fairly won."

Chapter Nineteen

The Western Theater: Hood's Tennessee Campaign

Following the loss of Atlanta to Sherman, Hood proposed that he take the Army of Tennessee west to strike Sherman's communications. Apparently without a clear plan or objective, initially just trying to pull Sherman back north to protect his line of communication, Hood then decided to move against Nashville. Sherman, apparently unconcerned, detached Thomas and elements of the XXIII and IV Corps under Major General Schofield to Nashville to deal with Hood while he continued with his plans for his famous "March to the Sea."

Hood crossed the Alabama-Tennessee border on November 21 and on November 27 made contact with Schofield and about 30,000 Federals at Columbia. Hood flanked the Columbia position and stole a march on Schofield, who was somewhat casual about his retreat. In a still unclear sequence of events involving misunderstood orders and just plain bad luck, the Army of Tennessee was in position to cut Schofield off from the main pike leading to Nashville but allowed the whole

Federal force literally to march past their campfires on the night of November 29 at Spring Hill.

Schofield, appreciating his narrow escape, moved to the previously prepared Federal defensive works at Franklin, on the Harpeth River. Not intending to offer Hood battle at Franklin, Schofield nevertheless needed time to repair the railroad bridge over the river so his trains could be moved. Hood for his part, was furious at his lost opportunity at Spring Hill and ordered a series of frontal assaults with only two of the three corps of his army present and practically no artillery support.

Hood wrecked his army at Franklin. He lost at least 1,750 men killed, 3,800 wounded, and 702 captured for a total of 6,252 compared to 2,326 Federal casualties. Among Hood's losses were five generals killed (including Patrick Cleburne), one captured, and six wounded (other accounts say that Hood probably lost at least 7,500 men at Franklin). His command structure was a shambles. Many regiments were now no larger than companies had been earlier in the war. In some cases the highest surviving ranking officer was a captain.

The remnants of the Army of Tennessee would follow the Federals to Nashville, but when Thomas had finally gathered his forces together and planned his attack, it was no contest.

On December 15-16, Thomas forced back and then routed Hood's remaining forces. Only a brilliant rear-guard action conducted initially by Stephen Lee's corps and later by Major General Nathan Bedford Forrest saved the army from complete destruction during its retreat.

When the surviving fragments of the Army of Tennessee went into winter quarters at Tupelo, Mississippi in January, 1865, Hood was relieved of command. The last Confederate offensive campaign of the war had ended.

Chapter Twenty

The Western Theater: Sherman's March to the Sea and
Campaign of the Carolinas

On November 12, 1864, Sherman marched out of Atlanta toward the
Atlantic coast. Tracing a line of march between Macon and Augusta, he
carved a sixty-mile wide swath of destruction in the Confederacy's
heartland. The only forces the Confederacy could bring to oppose him
was Wheeler's cavalry and a motley collection of militia and over and
under-aged reserves of perhaps 14,000 troops; certainly no match for
the 62,000 Union veterans Sherman had kept with him upon leaving
Atlanta.

His army marched in two large columns under the command of
Howard and Slocum. Sherman reached Savannah on December 10.
The Confederate garrison could not hope to prevent its capture, so
evacuated the city with 10,000 troops via a pontoon bridge. Sherman
presented Savannah to Lincoln as a "Christmas gift."

Sherman did not linger long at Savannah, and despite the miserable winter weather was soon on the march again. The Confederate forces in the region were fragmented at this time, with troop concentrations under Hardee and Beauregard, who could do little with the forces either had at hand, to slow Sherman down.

Columbia, South Carolina, captured on February 17, 1865, was dealt with particularly harshly by Sherman's men. Two-thirds of the city was burned down, although it was probably done at their own initiative rather than under any orders from Sherman. Many Federal troops held a special hatred for South Carolina because they felt the state was responsible for starting the war.

Finally, too late to really make any difference, Robert E. Lee was named General-in-Chief of the Confederacy's armed forces and Joe Johnston was given command of all remaining forces in North Carolina. Reinforcements from the tattered remnants of the Army of Tennessee would arrive via a patchwork railroad/overland route from Tupelo to join other commands under Beauregard, Bragg, and Hardee, but these were too few and too late.

Johnston looked for an opportunity to do some damage to Sherman's Federal steamroller and finally saw an opportunity on March 19, 1865. Slocum's and Howard's columns had become widely separated and Johnston concentrated his available troops (about 21,000 effectives) near Bentonville to try and crush Slocum's column before Howard could come to his support.

Initially the Confederate attacks went well, but Slocum was able to bring up reinforcements to withstand the repeated assaults. Little fighting took place on March 20, but on the 21st Sherman's entire command was in position to launch a counterattack. Johnston skillfully beat back the Federal attacks and retreated that night toward Smithfield.

Chapter Twenty-One

The Eastern Theater: Petersburg and Appomattox

Inside the Richmond fortifications, Lee's army had slowly been starving. Morale began to take a nose dive and between mid-February and mid-March about 8% of the men deserted. Grant meanwhile, took little offensive action but continued to extend his line southward which served to stretch Lee's already over-extended defenses almost to the breaking point.

On March 25, 1865, Lee attempted to break through Grant's lines at Fort Stedman, but despite some initial success, was finally beaten back with heavy losses. On April 1, Sheridan broke the Southside Railroad, defeating Pickett's division in the Battle of Five Forks. Sheridan was now in Lee's rear and the Danville Railroad, Lee's last supply line, lay fully exposed to capture. Lee had no choice. He ordered the immediate evacuation of Richmond. The Confederate government and the remaining gold in its treasury were put on a special train and headed south. Richmond fell on April 2.

Lee's last gamble was against long odds, but he had beaten long odds before and he was determined to try one last time. He had to try and link his army up with the remaining forces Joe Johnston had under his command. Perhaps if they could combine forces quickly enough, they could turn on either Grant or Sherman before the other could come to his aid. However, at this point in time, it would be problematic, even if combined, that the Confederate forces could defeat either Grant or Sherman alone.

To implement his plan, Lee set off along the line of the Appomattox River, seeking a point at which he could turn south. Grant however, kept moving his own army in a parallel course and prevented Lee from changing direction.

Finally, by the time Lee reached Appomattox Court House, and not finding rations that were supposed to be waiting for him, saw that Grant had succeeded in cutting off any further retreat. On April 9, 1865, he surrendered the Army of Northern Virginia to Grant.

Chapter Twenty-Two

The Western Theater: The End

Meanwhile Johnston, upon learning of Lee's surrender, asked Sherman for terms and formally surrendered his forces on April 26, 1865, at Durham Station, North Carolina. Other Confederate resistance came to an end following additional Federal offensives by Wilson's cavalry corps into the Alabama heartland and Canby's operations against Mobile.

The Trans-Mississippi Confederates also lay down their arms in May and June. The last Confederate general to capitulate, Brigadier General Stand Watie, leader of the Confederate Cherokees, surrendered to Federal forces on June 23, 1865. After four long and bloody years, the American Civil War had finally ended.

About the Author

Why my interest in the Civil War? I suppose that would only be natural for someone who, as a kid, grew up only 30-miles from Shiloh National Military Park. In 1993 however, I discovered my maternal great-grandfather was a member of the 2nd Mississippi Infantry Regiment. That discovery is what

Michael R. Brasher

finally "got the ball rolling" for me in a serious way. I am still working on a regimental history by the way, so I will keep you posted on my progress.

I was born and raised in West Tennessee, near where my great-grandfather moved following the Civil War. I served 20 years in both an enlisted and officer capacity in the United States Air Force. At the present time I still work in the defense industry as a systems engineer.

I graduated from the University of Tennessee with a degree in Electrical Engineering. While still in the Air Force, I obtained an MBA from Baldwin-Wallace College. Later, after retiring and vowing not to allow any of my GI Bill educational benefits to go unused, I obtained a MA in history (with a specialty in Civil War Studies) in 1999 from American Military University.

Please contact me at mrbrasher@yahoo.com to be added to my mailing list.

Also please visit my website at mrbrasher.com (still under construction as of 5/28/2018...please be patient) for additional information about upcoming releases.

facebook.com/Coonewah.Creek.Publishing

twitter.com/CoonewahP